Microgrids

Big Data for Industry 4.0: Challenges and Applications
Series Editors: Sandhya Makkar, K. Martin Sagayam, and Rohail Hassan

Industry 4.0, or the fourth industrial revolution, refers to interconnectivity, automation and real-time data exchange between machines and processes. There is tremendous growth in big data from the Internet of Things (IoT) and information services which drives the industry to develop new models and distributed tools to handle big data. Cutting-edge digital technologies are being harnessed to optimize and automate production, including upstream supply-chain processes, warehouse management systems, automated guided vehicles, drones and so on. The ultimate goal of Industry 4.0 is to drive manufacturing or services in a progressive way to be faster, effective and efficient, which can only be achieved by embedding modern-day technology in machines, components and parts that will transmit real-time data to networked IT systems. These, in turn, apply advanced soft computing paradigms such as machine learning algorithms to run the process automatically without any manual operations.

The new book series will provide readers with an overview of the state-of-the-art in the field of Industry 4.0 and related research advancements. The books will identify and discuss new dimensions of both risk factors and success factors, along with performance metrics that can be employed in future research work. The series will also discuss a number of real-time issues, problems and applications with corresponding solutions and suggestions, sharing new theoretical findings, tools and techniques for Industry 4.0 and covering both theoretical and application-oriented approaches. The book series will offer a valuable asset for newcomers to the field and practicing professionals alike. The focus is to collate the recent advances in the field so that undergraduate and postgraduate students, researchers, academicians and industry people can easily understand the implications and applications of the field.

Industry 4.0 Interoperability, Analytics, Security, and Case Studies
Edited by G. Rajesh, X. Mercilin Raajini, and Hien Dang

Big Data and Artificial Intelligence for Healthcare Applications
Edited by Ankur Saxena, Nicolas Brault, and Shazia Rashid

Machine Learning and Deep Learning Techniques in Wireless and Mobile Networking Systems
Edited by K. Suganthi, R. Karthik, G. Rajesh, and Ho Chiung Ching

Big Data for Entrepreneurship and Sustainable Development
Edited by Mohammed el Amine Abdelli, Wissem Ajili Ben Youssef, Ugur Ozgoker, and Imen Ben Slimene

Entrepreneurship and Big Data
The Digital Revolution
Edited by Meghna Chhabra, Rohail Hassan, and Amjad Shamim

Microgrids
Design, Challenges, and Prospects
Edited by Ghous Bakhsh Narejo, Biswaranjan Acharya, Ranjit Singh Sarban Singh, and Fatma Newagy

For more information on this series, please visit: www.routledge.com/Big-Data-for-Industry-4.0-Challenges-and-Applications/book-series/CRCBDICA

Microgrids

Design, Challenges, and Prospects

Edited by
Ghous Bakhsh Narejo, Biswaranjan Acharya,
Ranjit Singh Sarban Singh, and Fatma Newagy

CRC Press is an imprint of the
Taylor & Francis Group, an **informa** business

MATLAB® is a trademark of The MathWorks, Inc. and is used with permission. The MathWorks does not warrant the accuracy of the text or exercises in this book. This book's use or discussion of MATLAB® software or related products does not constitute endorsement or sponsorship by The MathWorks of a particular pedagogical approach or particular use of the MATLAB® software.

First edition published 2022
by CRC Press
6000 Broken Sound Parkway NW, Suite 300, Boca Raton, FL 33487–2742

and by CRC Press
2 Park Square, Milton Park, Abingdon, Oxon, OX14 4RN

© 2022 Taylor & Francis Group, LLC

CRC Press is an imprint of Taylor & Francis Group, LLC

Reasonable efforts have been made to publish reliable data and information, but the author and publisher cannot assume responsibility for the validity of all materials or the consequences of their use. The authors and publishers have attempted to trace the copyright holders of all material reproduced in this publication and apologize to copyright holders if permission to publish in this form has not been obtained. If any copyright material has not been acknowledged please write and let us know so we may rectify in any future reprint.

Except as permitted under U.S. Copyright Law, no part of this book may be reprinted, reproduced, transmitted, or utilized in any form by any electronic, mechanical, or other means, now known or hereafter invented, including photocopying, microfilming, and recording, or in any information storage or retrieval system, without written permission from the publishers.

For permission to photocopy or use material electronically from this work, access www.copyright.com or contact the Copyright Clearance Center, Inc. (CCC), 222 Rosewood Drive, Danvers, MA 01923, 978–750–8400. For works that are not available on CCC please contact mpkbookspermissions@tandf.co.uk

Trademark notice: Product or corporate names may be trademarks or registered trademarks and are used only for identification and explanation without intent to infringe.

Library of Congress Cataloging-in-Publication Data

Names: Narejo, Ghous Bakhsh, editor.
Title: Microgrids : design, challenges, and prospects / edited by Ghous Bakhsh Narejo, Biswaranjan Acharya, Ranjit Singh Sarban Singh, and Fatma Newagy.
Other titles: Microgrids (CRC Press)
Description: First edition. | Boca Raton : CRC Press, 2022. | Series: Big data for industry 4.0 : challenges and applications | Includes bibliographical references and index.
Identifiers: LCCN 2021018047 (print) | LCCN 2021018048 (ebook) | ISBN 9780367487959 (hbk) | ISBN 9780367639846 (pbk) | ISBN 9781003121626 (ebk)
Subjects: LCSH: Microgrids (Smart power grids)
Classification: LCC TK3105 .M5524 2022 (print) | LCC TK3105 (ebook) | DDC 621.31—dc23
LC record available at https://lccn.loc.gov/2021018047
LC ebook record available at https://lccn.loc.gov/2021018048

ISBN: 978-0-367-48795-9 (hbk)
ISBN: 978-0-367-63984-6 (pbk)
ISBN: 978-1-003-12162-6 (ebk)

DOI: 10.1201/9781003121626

Typeset in Times
by Apex CoVantage, LLC

Contents

Preface vii
Editors ix

Chapter 1 Microgrid Design and Challenges for Remote Communities 1

Fawad Azeem, Ghous Bakhsh Narejo, Waleed Rafique, Aizaz Mohiuddin, and Tauseef Anwar

Chapter 2 Microgrid Design Evolution and Architecture 19

Kumari Namrata, Ch Sekhar, D.P. Kothari, and Sriparna Das

Chapter 3 Renewable Energy Source Design in Microgrids 53

Kumari Namrata, Sriparna Das, R.P. Saini, and Ch Sekhar

Chapter 4 Microgrid Power System Control Designs 83

Renuka Loka, Alivelu M. Parimi, and P. Shambhu Prasad

Chapter 5 Hybrid Microgrid Design Based on Environment, Reliability, and Economic Aspects 101

Sriparna Roy Ghatak, Aashish Kumar Bohre, and Parimal Acharjee

Chapter 6 Trends in Microgrid Control 119

Anup Kumar Nanda, Babita Panda, Chinmoy Kumar Panigrahi, Arjyadhara Pradhan, and Naeem Hannoon

Chapter 7 Utility Tariff Variation and Demand Response Events: A Case Study of Microgrid Design and Analysis 137

Aashish Kumar Bohre and Parimal Acharjee

Chapter 8 SOS-Based Load Frequency Controller Design for Microgrids Using Degrees-of-Freedom PID Controller 181

Sunita Pahadasingh, Chitralekha Jena, and Chinmoy Kumar Panigrahi

Chapter 9 Performance Improvement of a UPQC Integrated with a Microgrid System Using Modified SRF Technique 201

Sarita Samal, Prasanta Kumar Barik, and Prakash Kumar Hota

Chapter 10 Application of Probabilistic Neural Network and Wavelet Analysis to Classify Power Quality 217

Pampa Sinha and Chitralekha Jena

Chapter 11 Overview of Security and Protection Techniques for Microgrids 231

Asik Rahaman Jamader, Puja Das, Biswaranjan Acharya, and Yu-Chen Hu

Chapter 12 Multilevel Voltage-Based Coordinating Controller Modeling, Development, and Performance Analysis 255

Ranjit Singh Sarban Singh, Tiara Natasya Abdul Halim, T. Joseph Sahaya Anand, and Maysam Abbod

Chapter 13 Case Studies on Microgrid Design Using Renewable Energy Sources 279

Arjyadhara Pradhan, Babita Panda, and Rao Mannepalli

Index 301

Preface

Microgrid is a technology that has been introduced to produce self-sufficient and sustainable energy systems. Within the microgrid, there is a diversity of distributed energy systems, such as solar panel systems, wind turbine systems, and combined heat and power systems, as well as many other types of generators that produce power for utilization. Looking at the diversity advantages of microgrids, as well as the growing number of power consumers opting for renewable energy sources, the evaluation of power systems—especially the microgrid type of power system—has become challenging and essential for development improvement. In addition, the growth of emerging power infrastructures opting into designed microgrids has resulted in extensively extending microgrid research and development, not only locally but also internationally. With so much improvement to look at, this book discusses design and development techniques, smart control and switching mechanisms, and innovations that can be used in the future in research or development of microgrids. Furthermore, the combination of diverse research techniques presented in this book will allow for the exploration of new findings involving multidisciplinary approaches in microgrids. With this exploration, it will offer great motivation for authors and editors to bring forth a book that fills the necessary gap as well as introduces new technological microgrid design and development. This text is an international forum for comprehensive researched content matter for design, development, fundamental theories and principles, and control and switching techniques, as well as coordination and management of the proposed power system applications in the environment, energy, electronics, and electric sectors. Researchers, engineers, students, academicians, and scientists of all levels involved in design and development of microgrid power systems will benefit from this book. Readers will gain an understanding of the practical and highly researched area of microgrids for their specific disciplinary areas or fields. The goal of this book is to provide readers with a broad review and presentation of all designs and development, which are highly sensitive findings for microgrid disciplinary research. Therefore, this text provides readers with the entire gamut of smart and intelligent designs, developments, fundamental theories and principles, and control and switching techniques, as well as coordination and management of microgrids. Also, the vision of this book is for the future and moves beyond existing microgrid designs and development to explore new realms of microgrid designs, developments, and applications.

Dr. Ghous Bakhsh Narejo
Associate Professor, Department of Electronic Engineering,
Faculty of Electrical, Computer Engineering,
NED University of Engineering & Technology, Pakistan

Biswaranjan Acharya
Faculty, School of Computer Engineering,
KIIT Deemed to be University,
Bhubaneswar, Odisha, India

Ir. Ts. Dr. Ranjit Singh Sarban Singh
*Senior Lecturer, Advanced Sensors and
Embedded Control (ASECs) Research Group,
Centre of Telecommunication Research Innovation (CeTRI),
Faculty of Electronic and Computer Engineering (FKEKK),
Universiti Teknikal Malaysia Melaka (UTeM), Hang Tuah Jaya,
Melaka, Malaysia*

Dr. Fatma Newagy
*Associate Professor, Electronics Engineering and Electrical
Communications, Ain Shams University, Egypt*

Editors

Ghous Bakhsh Narejo, PhD, is an associate professor in the Department of Electronic Engineering in the Faculty of Electrical, Computer Engineering at NED University of Engineering & Technology in Pakistan. He has 20 years of academic experience in electronic engineering and teaches undergraduate, master's, and PhD courses related to electronic engineering, microsystem design, industrial electronics, renewable energy, and imaging and analog ICs. He earned his master's degree in microsystems and electronic engineering from NED University, and his PhD in electrical engineering, specializing in nano devices, from Michigan Technological University, USA. His research is in microgrid design. Dr. Ghous has served in the Electronic Engineering Department at NED University, Karachi, Pakistan for the last 20 years. He supervises multiple PhD students, three of whom have successfully completed their PhDs at NED University. Dr. Ghous has to his credit numerous research papers. He has authored many research articles published in international journals of repute.

Biswaranjan Acharya, PhD, is currently associated with the Kalinga Institute of Industrial Technology (deemed to be university) and is pursuing a PhD in computer application from Veer Surendra Sai University of Technology (VSSUT), Burla, Odisha, India. He received an MCA in 2009 from IGNOU, New Delhi, India and an MTech in computer science and engineering in 2012 from Biju Pattanaik University of Technology (BPUT), Odisha, India. He is associated with various educational and research societies like IEEE, IACSIT, CSI, IAENG, and ISC. With his two years of industry experience as a software engineer, he has a total of ten years' experience in both academia of reputed universities like Ravenshaw University and in the software development field. He is currently working in the research areas of multiprocessor scheduling in different fields like data analytics, computer vision, machine learning, and IoT. He has 14 patents in his name and another 24 published to his credit. He has published research articles in internationally reputed journals and has served as a reviewer in many peer-reviewed journals.

Ranjit Singh Sarban Singh, PhD, is a senior lecturer in the Department of Computer Engineering in the Faculty of Electronic and Computer Engineering, Universiti Teknikal Malaysia Melaka (UTeM). He served as a tutor from 2008–2010, lecturer from 2010–2015, and is presently a senior lecturer since 2015. He is also attached to the Advanced Sensors & Embedded Control (ASECs) Research Group, Centre of Telecommunication, Research & Innovation (CeTRI) for his research activities. He received his PhD in general control system engineering (renewable energy system management) from Brunel University London, UK. He received his MEngSc in computer engineering (research) from Multimedia University Melaka, Malaysia in 2010. He received his BEng (honours) in electronics engineering (computer engineering) from Kolej Universiti Teknikal Malaysia Melaka (KUTKM), Melaka, Malaysia in 2006. During his service at UTeM from 2008–present, he taught numerous computer engineering subjects such as microprocessor technology, engineering mathematics, microcontroller technology, embedded system design, digital systems, principles of electrical and electronics, and other computer engineering-related subjects. He received his diploma in computer technology from Politeknik Seberang Perai (PSP), Pulau Pinang, Malaysia in 2004. Dr. Ranjit Singh is registered as a Professional Engineer (PE) with the Board of Engineers Malaysia (BEM). He is also registered as a Chartered Engineer of IET (UK), IntPE (UK), and a Professional Engineer (UK), as well as a Professional Technologist (Ts), with the Malaysia Board of Technologists (MBOT).

His research interests span several different disciplines: renewable energy system management, battery energy storage management, embedded system design, energy management algorithms, and consumer electronics system design. He has actively pursued research in these fields and published more than 60 journal and conference papers, filed 5 patents nationally, and been granted 1 trademark nationally. He has organized and chaired international conferences and continues to serve as a member of several scientific and organizing committees of international journals and conferences. He continues to review journal and conference papers. He has been a principal researcher for numerous research grants nationally and at universities. He has also won more than ten national and international innovation and invention awards.

Dr. Fatma Newagy, PhD, graduated from the Faculty of Engineering, Cairo University, where she worked as a research assistant and obtained a master's and PhD in electronics and communications engineering. She worked as an assistant professor in Cairo University (2008–2011) until she moved to Ain Shams University to be assistant professor (2011–2016) and associate professor (2017–present). Dr. Newagy is a member of the Space Research and Remote Sensing Scientific Council, Specialized Scientific Councils–Egyptian Academy of

Scientific Research and Technology (SSC-ASRT). She is a steering committee member at the MIT–ASU Center of Excellence–Energy founded by USAID. She is also a member of the Arab Foundation of Young Scientists.

Dr. Newagy has more than 20 years in total academic and research experience in many government and international universities in Egypt. She is an editorial board member on *Ain Shams Engineering Journal (ASEJ)*, Elsevier, and *SCIREA Journal of Electronics & Communication*, Science Research Association. She is a peer reviewer for many international journals with high impact factors. Dr. Newagy has participated in many research projects and grants. She is the supervisor of Space Innovation Lab in ASU, funded by the Egyptian Space Agency (EgSA). She has more than 60 international research papers, all in international journals with high impact factors, and has participated in international conferences worldwide in wireless communications, smart grids, energy efficiency, satellite communications, IoT, and innovations.

1 Microgrid Design and Challenges for Remote Communities

Fawad Azeem, Ghous Bakhsh Narejo, Waleed Rafique, Aizaz Mohiuddin, and Tauseef Anwar

CONTENTS

1.1 Introduction to Islanded Microgrids 2
1.2 Remote Communities and Potential of Microgrids 2
1.3 Current Trends in Design and Architecture of Islanded Microgrids 3
1.4 Different Types of Microgrids and Associated Loads 4
    1.4.1 Residential Microgrids 4
    1.4.2 Commercial Microgrids 4
    1.4.3 Special-Purpose Microgrids 5
1.5 Isolated Remote Microgrid Load and Operational Characteristics 5
1.6 Design Considerations of Islanded Microgrids for Remote Communities 6
    1.6.1 Residential Load Model for Remote Communities 7
    1.6.2 Commercial Load Model for Remote Communities 7
    1.6.3 Agricultural Load Model for Remote Communities 8
    1.6.4 PV Model 8
    1.6.5 Wind Turbine Mathematical Model 8
    1.6.6 Battery System 9
    1.6.7 Transmission System 9
1.7 Islanded Microgrid Management and Control 9
    1.7.1 Forecasting Techniques 10
    1.7.2 Demand Forecasting 11
        1.7.2.1 Long-Term Forecasting 12
        1.7.2.2 Short-Term Forecasting 12
    1.7.3 Generation Forecasting 12
1.8 Communication Techniques 13
    1.8.1 Communication Methods and Technologies Available for Microgrid Systems 13
        1.8.1.1 Power Line Communication 13
        1.8.1.2 Zig-Bee 13
        1.8.1.3 Cellular Network Communication 13

DOI: 10.1201/9781003121626-1

1.9 AC/DC Inverters and Power Electronics Used in Islanded Microgrids 14
1.10 Optimal Sizing of Islanded Microgrids 14
   1.10.1 Single Algorithm 14
   1.10.2 Classical Technique 15
1.11 Economics of Microgrids and Revenue Generation Methodologies 15
   1.11.1 Monthly Payment Plans 15
1.12 Isolated Community Microgrid Challenges 15
   1.12.1 Technical Challenges 16
   1.12.2 Social Challenges 16
1.13 Future Prospects 16
References 17

1.1 INTRODUCTION TO ISLANDED MICROGRIDS

Islanded microgrids, as the name implies, are isolated systems responsible for self-generation, transmission and distribution [1]. Microgrids are low-voltage systems with a defined load near a generation system. There are two major types of microgrids: islanded and grid-connected networks. A grid-connected microgrid is linked with the mainstream network via a point of common coupling (PCC), where islanded systems are self-sustainable systems. Renewable energy–based distributed generation is mostly used in recent microgrids for production of cleaner energy and to satisfy growing environmental concerns. With increasing penetration of renewable energy–based distributed generation sources, issues related to power quality and stability need special attention. Voltage, current and frequency deviation issues are well addressed in grid-connected microgrids, whereas islanded microgrids are responsible for self-control, which makes it more challenging compared to grid-connected systems. To address stability issues, battery storage devices, diesel generators and other storage devices such as super capacitors and so on are used. However, battery storage devices are mostly used for power conditioning and stability control.

The major component in an islanded microgrid system consists of control of the microgrid. The controller is responsible for ensuring the stability and economic operation of microgrids while controlling the power dispatch from the battery, distributed generation sources, load management and optimal power flow during the night hours or during undesirable events such as rainy/cloudy days or any event that reduces power output from distributed generation sources. Figure 1.1 shows major components of an islanded microgrid.

1.2 REMOTE COMMUNITIES AND POTENTIAL OF MICROGRIDS

Nearly one billion people live without electricity [1–3]. The living standard in such communities is low and lacks several basic necessities such as state-of-the-art hospitals, schools and other requirements of daily life. However, microgrids equipped with green energy resources serve as potential candidates to invest in rural electrification [4]. Rural electrification using intermittent distributed generators has shown

Microgrid Design for Remote Communities

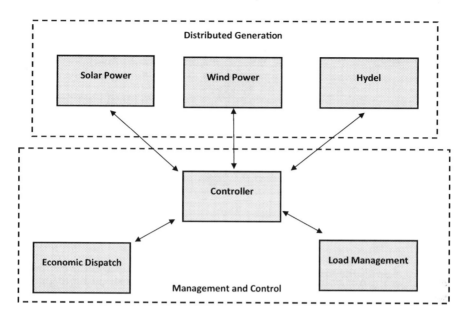

FIGURE 1.1 Islanded microgrid major components.

good potential for electrifying unprivileged areas with green energy. There is considerable growth in installation of the off-grid systems using solar and wind energy in developing countries. These systems help run residential areas, small businesses such as shops, small processing plants and community places such as hospitals, places of worship and schools. However, the intermittent nature of major sources of generation like solar and wind energy causes challenges such as severe power quality issues like voltage sags and frequency issues that eventually lead to blackouts [5, 6]. There is ongoing research to provide quality power to renewable energy–based islanded microgrids using different configurations, control algorithms, energy and demand management and utilization of storage systems for economical and reliable operation of islanded communities. The structure of the microgrids plays an important role in the reliable operation of microgrids and heavily depends on the nature of the load that runs in the main target area. The next section discusses the design and architecture of microgrid implementation in remote communities on the basis of their loads and applications.

1.3 CURRENT TRENDS IN DESIGN AND ARCHITECTURE OF ISLANDED MICROGRIDS

The basic architecture of an islanded microgrid is given in Figure 1.2. The architecture shows the major constituents of the microgrids, including solar photovoltaic (PV) distributed generators [1, 7, 17, 18], AC wind turbine generators, inverters connected with a DC PV power source and the whole generation source connected

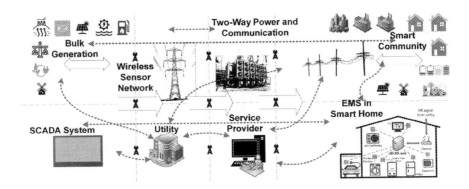

FIGURE 1.2 Microgrid architecture.

to an AC bus bar. To further enhance the reliability and power quality, the battery storage is configured to be charged from solar power and provide power to the load. The battery storage is responsible for frequency and voltage regulations and also provides power during the night hours. The load, storage and distributed generation is connected to the centralized controller. The centralized controller is the heart of the islanded microgrid and performs load management, economic and appropriate power dispatch from storage using feedback systems. The decisions are based on the available past, current and future information using forecasting, sensory information load and weather profiles of the area.

1.4 DIFFERENT TYPES OF MICROGRIDS AND ASSOCIATED LOADS

Based on the application and needs, microgrids can be of different types with different load applications. Power utilization with respect to loads is different between the various microgrids. Some common microgrids being utilized are given in the following.

1.4.1 Residential Microgrids

Residential microgrids are the most common type of microgrid being used to feed the domestic load. The most common loads for residential microgrids are fans, TVs, LEDs, refrigerators and other household loads. Isolated communities often exist in rural areas of developing nations that utilize load during the night hours. That is why load utilization characteristics are different and depend heavily on battery storage. The cost of battery storage increases the overall capital cost of the system.

1.4.2 Commercial Microgrids

Commercial microgrids are used to provide power to agricultural, industrial and other commercial loads such as shops and so on. The load in such microgrids usually operates during the daytime, while a minimal residential load runs during the night hours. However, the agricultural loads are heavy compared to residential microgrids.

Microgrid Design for Remote Communities

The operation and maintenance cost for such loads is comparatively lower than for residential microgrids due to lower battery storage requirements.

1.4.3 Special-Purpose Microgrids

Special-purpose microgrids are those that run remote military bases, campuses, hospitals and so on, where access to the mainstream network is not available. The loads are mostly similar to residential microgrids, including some special-purpose loads. The utilization of loads is mostly uniform during the day and night. To provide reliability and avoid blackouts, such microgrids uses diesel generators to provide backup power for sensitive loads.

1.5 ISOLATED REMOTE MICROGRID LOAD AND OPERATIONAL CHARACTERISTICS

Table 1.1 shows the operational characteristics of three types of microgrids and their utilization in isolated areas under different conditions. A residential microgrid is equipped with storage systems that increase the overall capital cost. Unlike remote residential microgrids, agricultural microgrids normally work during the daytime and utilize directly available power, in most cases solar energy to feed heavy loads; this reduces the cost for battery storage, which eventually reduces the capital cost of the system. Special-purpose microgrids use loads that run 24 hours and require storage as well as diesel generators in most cases for reliability. The capital cost for special-purpose microgrids is high compared to residential and commercial microgrids. Once residential microgrids are installed in rural isolated communities, the availability of skilled manpower is not guaranteed on permanent basis. The same is the case with commercial microgrids. On the other hand, special-purpose microgrids are organized with maintenance, fault detection and troubleshooting staff available within the microgrid system staff. The financial liabilities of commercial and residential microgrids are on the owners, whereas the government mostly takes responsibility for special-purpose microgrid financial liabilities. These liabilities include capital costs, operation and maintenance costs and other required costs.

TABLE 1.1
Operational Characteristics of Islanded Microgrids

Sr #	Characteristics	Residential Microgrids	Commercial Microgrids	Special-Purpose Microgrids
1	Load Operation	16–18 hours	8–10 hours	24 hours
2	Capital Cost	High	Low	High
3	Skilled Manpower Availability	Yes	No	Yes
4	Special-Purpose Loads	No	No	Yes
5	Financial Liabilities	Owner	Owner	Government
6	Social Benefits	Yes	Yes	Yes

1.6 DESIGN CONSIDERATIONS OF ISLANDED MICROGRIDS FOR REMOTE COMMUNITIES

Design of the microgrid is layered in three interconnected systems to perform operation in islanded mode. Figure 1.4 shows the layers of islanded microgrids in remote communities. These design layers support the controller in making effective decisions such as introducing load shedding, taking power from storage, charging storage and initiating demand management controlling responsive loads. All such actions are taken to reduce the frequency and voltage disturbances and to reduce potential blackouts. The layers are based on previous information available such as load and generation profiles and are considered permanent information. Permanent information is not adaptive and does not change with the passage of time. Permanent information may include the load profiles of the area, interviews with the community people and weather and climatic conditions of the area based on which distributed generators are selected. Permanent information helps in the initial design and appropriate system sizing.

The second layer is the adaptive layer that takes information from sensors, instantaneous load consumption, market prices, current weather information and availability of the state of the charge of storage.

The third layer is the decision and optimization layer that decides on the appropriate action based on the second layer. The operation of the third layer heavily depends on the information received from the second layer, that is, the adaptive layer. The decision is either made based on the availability of the adaptive data or the decision layer optimizes the values for better operation of the system.

In the following sections, the design parameters of each layer will be studied in detail for the appropriate operation of islanded microgrids in remote communities.

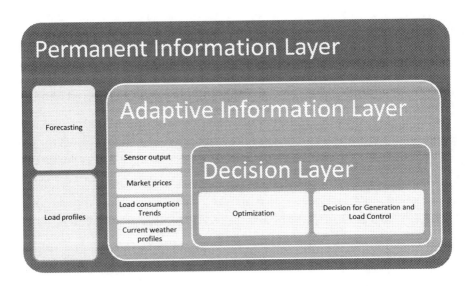

FIGURE 1.3 Layers of remote islanded microgrid.

Microgrid Design for Remote Communities

1.6.1 Residential Load Model for Remote Communities

The residential sector plays a vital role in energy demand. Currently, consumers constantly use more electrical appliances even in remote communities compared to the load trends in the last decades. It is due to an improved lifestyle. Domestic appliances can be categorized into brown goods and white goods. Small electronic consumer appliances are known as brown goods. Office and other communication equipment such as laptops, LCDs, printers, scanners, phones, routers and so on that are not part of small marketplaces in remote communities are also included in brown goods. White goods are also known as white-ware. This class has appliances like kettles, toasters, fans, irons, sewing machines, vacuum cleaners and so on that are used in daily routines such as cooking, cleaning and preservation [8]. However, the load contribution of white goods is still low in remote communities. Lighting also play crucial role in energy consumption. In the residential load model for the isolated communities, the load types that have considerable and continuous usage in the daily routine are taken into account. Equation (1.1) shows the residential loads.

$$L_R(t) = \left(P_{(t)}\right)\left(\sum_1^n P_{TV} + \sum_1^n P_{Ref} + \sum_1^n P_L + \sum_1^n P_F\right) \quad (1.1)$$

where $P_{(t)}$ is the instantaneous load running at time t and $\sum_1^n P_{TV}$ is a summation of the television load. $\sum_1^n P_{Ref}$ denotes the refrigeration load, $\sum_1^n P_L$ shows the lighting load and $\sum_1^n P_F$ shows the load of fans.

1.6.2 Commercial Load Model for Remote Communities

The commercial sector covers electrical loads that are mainly used in restaurants, shopping centers, malls and so on. Load consumption is higher during the day as compared to the residential load. Commercial loads usually have non-linear consumption patterns. These loads may produce power quality issues such as harmonics, voltage dips and frequency deviations. The commercial loads in remote communities are mainly made up of fans, laundry and lighting loads with some heavy water pump operation for irrigation purposes. Equation (1.2) shows commercial loads for remote communities.

$$L_{Com}(t) = \left(P_{(t)}\right)\left(\sum_1^n P_{Wp} + \sum_1^n P_{WM} + \sum_1^n P_L + \sum_1^n P_F\right) \quad (1.2)$$

where $\sum_1^n P_{wp}$, $\sum_1^n P_{WM}$, $\sum_1^n P_L + \sum_1^n P_F$ refer to the water pump, washing machine, light and cooling (fans) loads.

1.6.3 Agricultural Load Model for Remote Communities

Agricultural load includes activities that are mostly related to irrigation. Motors in agriculture are mainly used for water pumping. Agricultural loads may vary with the seasons (environment, weather change), for example, during a rainy day, there will be no or very low utilization of solar pumps to draw power for irrigation. The power required by agricultural sector is mainly used for activities such as planting, pumping, cultivating and so on. The inductive nature of water pumps may affect power stability during the operation of other parallel loads and hence may affect power quality. Switched capacitors and voltage regulators are often used to solve power quality issues [9].

1.6.4 PV Model

A PV cell is the basic unit of a PV system. Many PV cells combined together form a module. To get the required voltage, many modules are connected in series and parallel. PV cells convert sunlight into electrical energy. The output current from the equivalent model is described by Equation (1.4).

$$Ipv = Ncell\left[Isc - Io\left(e^{\left(\frac{Vpv - IpvRs}{Vt}\right)} - 1\right)\right] - \frac{Vpv - RsIpv}{Rp} \quad (1.4)$$

In the equation, as discussed in [10], I_{sc} represents a short circuit current under the given irradiance, the diode current is represented as *Id*, series resistance is *Rs* and *Io* is the diode saturation current. PV voltage is changed by switching the special design DC-DC converter to get the maximum average power output of the array [10].

1.6.5 Wind Turbine Mathematical Model

Wind turbines generate mechanical torque from wind in the form of kinetic energy and deliver it to a rotor for mechanical output that turns the generator shaft Tm to produce power Pm [11, 16]. The power obtained from wind is given as in [11].

$$P_w = \frac{1}{2}C_p(\lambda) * \rho A V^3 \quad (1.5)$$

In this expression, ρ represents air density (kg/m³), wind speed is represented by *V*, and *A* denotes the area of the blade. λ represents the tip speed of the turbine [11]:

$$\lambda = \frac{\omega R}{V} \quad (1.6)$$

In Equation (1.6), ω represents the blade angular velocity, *R* represents the radius and *V* is the velocity of the wind.

Microgrid Design for Remote Communities

From the value of power, we can find the mechanical torque on the shaft as [11]:

$$T_m = \frac{P_m}{\omega} \tag{1.7}$$

where T_m is the mechanical torque, P_m is the power and ω is the blade angular velocity.

1.6.6 BATTERY SYSTEM

The mathematical model of the battery is given in Equation (1.8) [10].

$$Eb = Eb,o - K\frac{Qb}{Qb - Ib,t} + A.e^{(-B.Ib,t)} \tag{1.8}$$

where Eb is the open circuit voltage, Eb,o is the open circuit voltage under standard conditions, Qb is the battery charge capacity, A and B are the constants linked to exponential terms, t is the time and Ib is the time integral of the battery current [10].

1.6.7 TRANSMISSION SYSTEM

Distributed generators are connected to designated loads via a low-voltage transmission system. The reactive power compensation is done using the storage system, as given in Equation (1.9) [12], where ΔP, ΔQ control the changes in power (active and reactive) generation/consumption. On the other hand, storage systems are also used for active power delivery during undesirable events as well as during the night hours.

$$U2 \approx U1 + \frac{Rline[Pder - PL \pm \Delta Pcontrol] + Xline[Qder - QL \pm \Delta Qcontrol]}{U2} \tag{1.9}$$

1.7 ISLANDED MICROGRID MANAGEMENT AND CONTROL

Microgrid management and control lie in the second layer of isolated microgrids and can be considered the heart of reliable operation. Figure 1.4 shows the hierarchy of microgrid management and control. Information sharing between microgrid components and controller, sensory information gathering and appropriate actuation such as controlled power dispatch and load shedding decisions are meant to be made based on the operation of the second layer. The complete islanded microgrid operation is based on robust control and information being transmitted at a particular instant of time. Adaptive information along with forecasting results provide a basis for decisions based on the microgrid control objectives.

For instance, the control algorithm aims to monitor the load and perform load shedding, and parameters such as short-term load forecasting, load consumption and state of the charge of storage parameters may be taken into account. The same is the

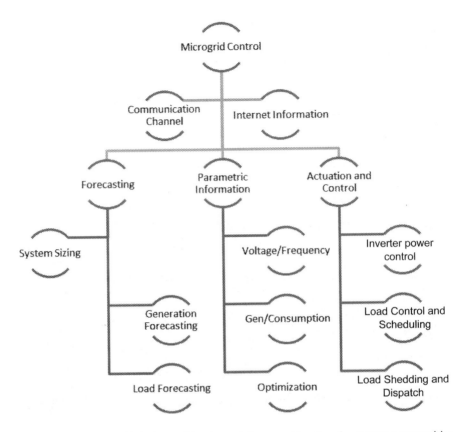

FIGURE 1.4 Microgrid design and implementation consideration for remote communities.

case with generation control, load scheduling, demand response and controlled power dispatch from storage. Frequency and voltage regulations are also termed control parameters and lie within the adaptive second layer of microgrid controls. Additional weather information is dependent on the availability of the internet. However, remote communities often do not have the option of quality internet service, especially in islanded rural areas of developing countries. Second-layer information is routed using communication channels and protocols to the third layer. The third layer is responsible for taking action based on the control decisions made. The third layer is the actuation and control layer and is responsible for obeying the decisions being made by the controller based on the second layer's information.

1.7.1 Forecasting Techniques

Forecasting lies in the first layer of microgrid design. Forecasting techniques are used to evaluate and analyze the demand and supply patterns under different conditions and events. Forecasting is essential for initial microgrid planning, especially for islanded microgrids. Figure 1.5 shows different types of forecasting techniques

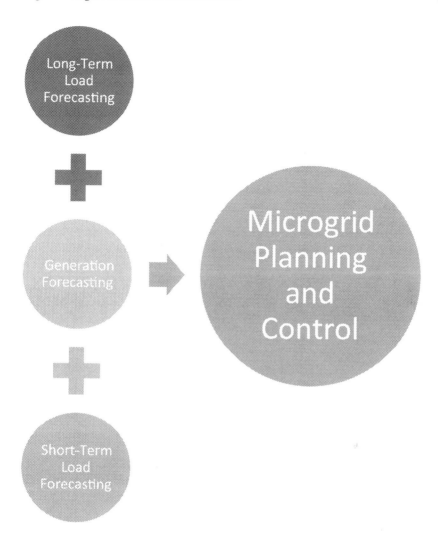

FIGURE 1.5 Types of forecasting used for microgrid planning and control.

used for microgrid implementation and planning. Formally, supply-side forecasting is known as generation forecasting, whereas demand-side forecasting is known as load forecasting. Several techniques are utilized for load and generation forecasting.

1.7.2 Demand Forecasting

Load forecasting is done to analyze the future load patterns of the community, which gives an idea of appropriate system sizing. Practically, accurate load estimation for future utilization gives appropriate overall system sizing. Load forecasting is also essential for financially viable systems. In isolated remote communities, the

TABLE 1.2
Parameters Required for Different Types of Forecasting

Parameters for Short-Term Forecasting	Parameters for Long-Term Forecasting	Forecasting Techniques
Load consumption	Load type	Regression analysis
Time of operation	Time of operation	Artificial intelligence
State of the charge of storage	Total power	Fuzzy logic
Available generations for current and next day	Time of day	Multi-criteria decision-making
Schedule of responsive loads	Holidays	Fuzzy-neuro hybrid forecasting
Market prices	Day of the month	

exact load consumption trends are often not available. In such cases, load trends of a nearby community having similar activities and residence style can be taken as a reference. On the other hand, having knowledge of some parameters such as types of load to be operational, their time of use and so on helps in estimating load trends. Load trends for isolated remote communities can also be gathered using interviews with residents.

There are two dominant forecasting types used: 1) long-term forecasting and 2) short-term forecasting. Both forecasting methodologies have different purposes.

1.7.2.1 Long-Term Forecasting

Long-term forecasting is used for microgrid planning purposes. The forecasting results are used to evaluate system sizing in the future using the parameters that affect the overall consumption of the system. Since load trends in remote communities are not much different and variations in consumption patterns are low compared to urban areas or highly dense areas, the parameters used for forecasting load trends can be trusted for overall microgrid planning and system sizing.

1.7.2.2 Short-Term Forecasting

Contrary to long-term forecasting, short-term forecasting methods are used especially for day-ahead load consumption trends. These trends are used to evaluate variations in the loads and adjust accordingly to get rid of blackouts or emergency conditions. Short-term forecasting is helpful in decision-making processes for controllers. Table 1.2 shows the parameters and techniques used in long- and short-term forecasting.

1.7.3 Generation Forecasting

Generation forecasting is used to predict the generation trends of distributed generation based on the climatic conditions of the area. It is essential to forecast the generation trends of the area for feasible microgrid operations. Generation forecasting is dependent on the availability of the climatic conditions of the area such as annual precipitation, number of cloudy days in the area, sunny days, average wind speed, wind gusts and availability of water resources in the case of hydel energy. Generation forecasting parameters are available on an accessible website that helps in evaluating the site while performing generation forecasting on a long-term basis. Generation forecasting is not only required for a long-term basis, but it also effectively helps in short-term planning such as day-ahead planning to decide on the operation of responsive loads, utilizing battery storage during day- or nighttimes and making decisions for load shedding.

1.8 COMMUNICATION TECHNIQUES

Communications systems play a vital role in microgrid (MG) systems. The communication system provides a connection with all units in the MG. The central controller (CC) must coordinate with all units to manage the microgrid. A local data concentrator is usually used to collect data from several groups of nodes and then send the data to the CC via a communication channel. Many technologies and channels for communication can be used in microgrids to ensure stability in the microgrid.

1.8.1 Communication Methods and Technologies Available for Microgrid Systems

A communication system is needed for power control, protection and restoration in a MG [13]. The basic methods of communication used in MGs are programmable logic controller (PLC), broadband over a power line, global systems for mobile communication, local-area network (LAN), wide-area network (WAN), TCP/IP, fiber optic, Wi-Fi 802.11b and Zig-Bee/IEEE 802.15.4.

1.8.1.1 Power Line Communication
PLC is a method that uses power lines to transfer high-speed (usually 2 to 3 Mb/s) data signals from one piece of equipment to the other. PLC uses power cables as communication media and today has received attention for smart power systems.

1.8.1.2 Zig-Bee
Zig-Bee is a wireless communications technology. The advantage is relatively low in usage of power, high data rate and complexity. It is a suitable technology for smart lightning systems, energy monitoring systems, home automation, automatic meter reading and so on.

1.8.1.3 Cellular Network Communication
Cellular networks can be the best option for communicating between direct grids and the utility. The existing communications system prevents utility operational costs and the additional time frame for making a strong communications infrastructure. 2G, 2.5G, 3G and Wi-MAX are the advanced technologies available for microgrid systems [14].

1.9 AC/DC INVERTERS AND POWER ELECTRONICS USED IN ISLANDED MICROGRIDS

The role of the smart inverter in islanded microgrids is to operate as an interface between power generation and consumption points. The basic function is not only to convert AC to DC or vice versa but also to control the power flow and to disconnect the system in case of any fault. The processing components of a MG are an inverter that collects data and automatically configures to help operate in a safe mode. The power converter acts as an interface for a distributed energy resource (DER) to the grid whose objectives are to fulfill the requirements of the IEEE 1547 standards series.

Grid-forming inverters have main objectives that are first grid connection and second islanded operation. In the case of grid connection, inverters control the power (active, reactive) injected into the AC bus to balance the state of charge for islanded operation. It is compulsory to generate voltage that is purely sinusoidal in the bus. Grid inverters in the form of voltage source inverters mainly help in this task.

In islanded mode, grid inverter control is typically changed to operate as a voltage-controlled source; these inverters can be also controlled to follow the grid. It also injects a given amount of current that helps grid-forming converters. In the case of grid-connected operation, the supporting inverters do not work or may be used to improve the power quality of the supply bus.

1.10 OPTIMAL SIZING OF ISLANDED MICROGRIDS

The planning and implementation phase for an islanded microgrid requires optimal sizing of the overall microgrid [15]. Optimal sizing is essential for the initial or capital investment cost of the microgrid that includes generation, transmission, distribution, installation and storage cost. It is most important to find out the appropriate sizes of renewable energy sources and the associated energy storage for reliable, efficient operation of the MG. Optimal sizing techniques can be categorized into classical or modern techniques. Classical methods use iterative, numerical, analytically and probabilistic and graphical construction techniques. These techniques use differential calculus to find the optimum solution. Modern methods use artificial methods. These techniques determine the global optimum system. A commonly used software tool in optimal sizing for standalone hybrid energy systems (HES) is the hybrid optimization model for electric renewables (HOMER). Other software, IHOGA, has been used in optimal sizing for standalone HES. Sizing optimization methods can use a single-objective optimization function and multi-objective optimization functions.

1.10.1 Single Algorithm

Single algorithms, which have classical and artificial techniques used to solve size optimization for photovoltaic-wind turbine hybrid energy systems, are reviewed in the following.

1.10.2 CLASSICAL TECHNIQUE

Rare studies have been carried out in classical methods for optimal sizing of microgrids. Algorithms such as DIRECT are an efficient technique used to find the global optimum of many problems. Some studies have been carried out in linear programming (LP) in optimizing the size of standalone HESs.

Mixed integer linear programming (MILP) takes into account the demand of energy at consumption points and maps to find the location and optimal size of the HES. The main function of optimization is to reduce the initial system cost that is used to find the performance of system. Studies found that the optimal location, in addition to optimal size, minimizes the initial costs.

1.11 ECONOMICS OF MICROGRIDS AND REVENUE GENERATION METHODOLOGIES

Islanded microgrid systems in remote communities need financial models to be deployed to build a self-sustainable system. Such communities cannot afford heavy investments, but the subsidy from the government is not realistic. This requires a unique financial revenue collection methodology for repair, maintenance and component replacement such as battery storage systems. The system to be deployed needs a revenue collection methodology that does not place a burden on the residents and serves the purpose of self-sustainable microgrid systems. A revenue generation methodology is developed in this chapter in the following.

1.11.1 MONTHLY PAYMENT PLANS

Monthly payment plans are designed on a per-capita basis. This means the number of people living in any house will have to pay the bill. The overall per capita can be evaluated based on the capital and recurring cost of the entire system divided by the total number of people living in that area.

$$X_c = M_C + R_c / P \tag{1.10}$$

where X_c is the per capita cost, M_C is the microgrid capital cost, R_c is the recurring maintenance cost and P is the number of people residing in the area.

For instance, if the total cost of the microgrid is 20,000 USD and the recurring maintenance cost is 500 USD/month and 6000 USD/year, and the number of people who benefit from the microgrid is 1000, then cost per capita will be 26 USD. Each household will have to pay 26 USD for a year to generate revenue of 26,000 USD per year and 2166 per month. Each participant will be bound to pay 26 USD a year, which is equivalent to 2.16 USD/month and is a nominal amount to pay easily.

1.12 ISOLATED COMMUNITY MICROGRID CHALLENGES

According to [18], isolated microgrid communities, especially in remote areas, face the challenge of 4As: accessibility, affordability, adaptability and applicability. The challenge lies in both technical and social issues for effective, sustainable, efficient and reliable microgrids. The technical and social challenges faced by the remote communities while implementing microgrids are given in the following.

1.12.1 Technical Challenges

- Research work on microgrids is mostly done at the lab scale under a controlled environment where the real ground issues are not addressed. Hence, most studies are based on the lab scale or simulations.
- There is a need to develop a robust communication infrastructure to deal with the control system that is the heart of the microgrid. In remote communities, internet to get real-time weather data is not available. On the other hand, sophisticated communication protocols are either established in urban areas or only at lab-scale implementation.
- There are no load generation trends available that are needed for optimal microgrid planning.
- The designed controller needs proper monitoring and sensory information; maintenance of the control system, sensors and communication system is an issue.
- Modern wireless communication, weather profiles and control systems are not accessible.

1.12.2 Social Challenges

- The cost of expensive microgrid components and generation systems is beyond the ability of the local people.
- Lack of education is the biggest hurdle, due to which local manpower cannot be deployed to maintain the control system and so on.
- The people do not accept change. Load consumption and scheduling is not properly followed.
- There is no technical manpower available to check if the system malfunctions.
- There is no proper revenue generation methodology that helps in self-sustainable system installation.

1.13 FUTURE PROSPECTS

The aging transmission and distribution network foundation is among the significant obstacles in planning a reliable and economically viable power system. The upgrade of the current transmission foundation needs huge investments, which are practically unrealistic. This circumstance has provided a reasonable opportunity for renewable power sources connected to a low transmission framework. Microgrids are developing all around the globe, especially in islanded areas that are not connected to mainstream networks. The development of the microgrid market has expanded exponentially since the year 2011 and is anticipated to reach an introduced limit of over 15 GW by 2022. Today, grounds or institutional microgrids are the biggest by application and are estimated to develop at a compound annual development rate (CAGR) of 18.83% from 2012–2022. It is obvious from the current situation and ideal conditions for microgrids that in the future, microgrids will fulfill unconnected areas' demands in developed as well as developing nations.

REFERENCES

1. Narejo, G. B., Azeem, F., & Ammar, M. Y. (2015, June). A survey of control strategies for implementation of optimized and reliable operation of renewable energy based microgrids in islanded mode. In *2015 Power Generation System and Renewable Energy Technologies (PGSRET)* (pp. 1–5). IEEE.
2. Azeem, F., Narejo, G. B., & Shah, U. A. (2018). Integration of renewable distributed generation with storage and demand side load management in rural islanded microgrid. *Energy Efficiency*, 1–19.
3. Azeem, F., & Narejo, G. B. (2020). A fuzzy based parametric monitoring and control algorithm for distinctive loads to enhance the stability in rural islanded microgrids. *Facta Universitatis, Series: Electronics and Energetics*, 33(2), 227–241.
4. Fowlie, M., Khaitan, Y., Wolfram, C., & Wolfson, D. (2018). *Solar microgrids and remote energy access: How weak incentives can undermine smart technology*. TechRep. Energy Institute at HAAS.
5. Blair, N., Pons, D., & Krumdieck, S. (2019). Electrification in remote communities: Assessing the value of electricity using a community action research approach in Kabakaburi, Guyana. *Sustainability*, 11(9), 2566.
6. Ustun, T. S., Ozansoy, C., & Zayegh, A. (2011). Recent developments in microgrids and example cases around the world—A review. *Renewable and Sustainable Energy Reviews*, 15(8), 4030–4041.
7. Nguyen, X. H., & Nguyen, M. P. (2015). Mathematical modeling of photovoltaic cell/module/arrays with tags in MATLAB/Simulink. *Environmental Systems Research*, 4(1), 24.
8. Dickert, J., & Schegner, P. (2010). Institute of electrical power systems and high voltage Engineering Technische Universität Dresden, Dresden, Germany. "Residential Load Model for Network Planning", Conference Paper Modern Electric Power Systems, Wroclaw, Poland.
9. Ali, F., Naeem Arbab, M., & Kashif Khan, M. (2019). Department of electrical engineering, US-PCASE UET Peshawar Pakistan, "SVC based voltage stabilization for sensitive agricultural loads". *Sarhad Journal of Agriculture*, 35.
10. Giorgio Caua, Daniele Coccoa, & Mario. (2014). Modelling and simulation of an isolated hybrid micro-grid with hydrogen production and storage. Department of Mechanical, Engineering, University of Cagliari, via Marengo 2, Conference Italian thermal Machine Associates, Elsevier.
11. Worku, M. Y., & Abido, M. A. (2017). PMSG based wind system for real-time maximum power generation and low voltage ride through. *Journal of Renewable and Sustainable Energy*, 9(1), 013304.
12. Krechel, T. S. F., Gonzalez, L., Chamorro, H., & Rueda, J. L. (2019). Transmission system-friendly microgrids an option to provide ancillary services. 2019 IEEE Milan Power Tech Italy, Elsevier, doi:10.1016/b978-0-12-817774-7.00011-9.
13. Mohagheghi, S., & Yang, F. (2011). Applications of microgrids in distribution system service restoration. ISGT, doi:10.1109/isgt.2011.5759139.
14. Mr. Chavan, P. D., & Prof. Devi, R. J. (2016). Survey of communication system for DG's and microgrid in electrical power grid. *International Research Journal of Engineering and Technology*, 3(7).

15. Al-falahi, M. D. A., Jayasinghe, S. D. G., & Enshaei, H. (2017). A review on recent size optimization methodologies for standalone solar and wind hybrid renewable energy system. *Energy Conversion and Management*. ISSN: 0196–8904.
16. Wang, C. N, Lin, W. C., & Le, X. K. (2014). Modelling of a PMSG wind turbine with autonomous control. *Mathematical Problems in Engineering*, doi:10.1155/2014/856173.
17. Patrao, I., Figueres, E., Garcerá, G., & González-Medina, R. (2015). Microgrid architectures for low voltage distributed generation. *Renewable and Sustainable Energy Reviews*, 43, 415–424.
18. Azeem, F., Narejo, G. B., & Shah, U. A. (2018). Integration of renewable distributed generation with storage and demand side load management in rural islanded microgrid. *Energy Efficiency*, 1–19.

2 Microgrid Design Evolution and Architecture

Kumari Namrata, Ch Sekhar, D.P. Kothari, and Sriparna Das

CONTENTS

- 2.1 Introduction 19
- 2.2 History of Microgrids 22
- 2.3 Need for Microgrids 23
- 2.4 Definitions of Microgrids 25
- 2.5 Structure of Microgrids 25
  - 2.5.1 Distributed Generation 26
    - 2.5.1.1 Stand-Alone Reciprocating Engine Generators 28
    - 2.5.1.2 Turbine Generators 28
    - 2.5.1.3 Solar PV Cells 29
    - 2.5.1.4 Wind Turbines 30
    - 2.5.1.5 Fuel Cells 30
  - 2.5.2 Energy Storage Technology 31
  - 2.5.3 Microgrid Loads 33
- 2.6 Operation of Microgrids 33
- 2.7 Monitoring of Microgrids 35
- 2.8 Stability, Control, and Strategies for Microgrids 36
  - 2.8.1 Microgrid Control Techniques 41
- 2.9 Interface Modules of Microgrids 41
- 2.10 Types of Microgrids 43
- 2.11 Pros and Cons of Microgrids 44
- 2.12 Microgrid Architecture 45
- 2.13 Commercial Planning of Microgrids 48
- 2.14 Conclusion 48
- References 48

2.1 INTRODUCTION

Human beings have been very fond of knowledge, innovative ideas, inventions and discoveries from the very beginning. One of the most successful and powerful

DOI: 10.1201/9781003121626-2

discoveries was the discovery of electricity through a lightning experiment conducted by Benjamin Franklin in the year 1752. Later, many experiments were conducted, and this led to invention of the light bulb by Thomas Edison and the AC generator by Nikola Tesla, and then the development of the power grid took place. The age of electricity in the United States began during the 1890s with the lighting of the Chicago World's Fair and completion of the first long-distance transmission line bringing hydro-electric power from the mighty Niagara Falls to the city of Buffalo [1]. The Industrial Revolution also supported the evolution of electricity by machines and other tools. The development of grids is still going on, which bought the introduction of the concept of the smart grid to twenty-first-century society. The microgrid plays a vital role in the formation of smart grids. In fact, without microgrids, the concept of building a smart grid is incomplete. So the study of microgrid concepts, design, evolution and architecture is important for an uninterruptible power supply to every customer connected to the main grid [2].

Knowing the meaning of electric grids is needed before discussing microgrids and smart grids. Most of us are well aware of grid systems. The electric grid or power grid is an interconnection of generating and distribution systems through transmission systems. The generation system consists of generating stations where electricity is generated and transmitted to the consumers in a distribution system. Distribution systems have different types of customers who require electric power. The transmission system consists of different levels of subsystems and transmission lines. These definitions are related to the present scenario where one-way communication is present, and a typical layout of an electrical power grid can be seen in Figure 2.1. One-way communication is the process of getting information related to problems in power utilization from the customer to the generating stations, but the customer is not well informed about the status of problem solving from the generating station. This problem can be solved by installing smart metering units and computers at both ends, and then two-way communication will be possible [3, 4].

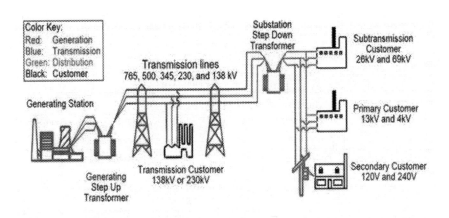

FIGURE 2.1 A typical layout of an electrical power grid.

Microgrid Design Evolution and Architecture

Until now, research models have been used to check the operation and efficiency of smart grids. The main challenge in electric power supply is to supply efficient power continuously without any interruptions and free from harmonics to the customer or industries. This problem can be minimized or solved by introducing a distributed generation unit near the customer or by the customer and connecting it to the main grid [5]. Distributed generation can be done by a standalone generator, a PV system, windmill system, fuel cell generators and so on. The concept of distributed generation units in the distribution system and connecting them to the main grid brings in the formation of microgrids. Distributed generation units or distributed energy resources can be defined as the devices used as electric supply, but these are not widely deployed in the main grid or in a centralized grid and are connected at distribution or lower voltages to meet the demand [6]. For example, PV arrays and wind turbines are used with battery storage so that electric power can be available even for non-emergency use like turning on lights, fans, and so on. Microgrids can be assumed to be a minor form of an actual grid because they are similar to a conventional grid with respect to the terms of power generation, transmission and distribution of power with control features [7, 8]. However, the technology of microgrids differs from technology of conventional grids with respect to the distance between power-generating sources and consumers, and it also differs with the consumption cycles because the installation of a microgrid is near the consumer or load sites. Table 2.1 shows a comparison between microgrids and conventional grids.

TABLE 2.1
Microgrids versus Conventional Grids

	Microgrid	Conventional grid
Generation	Micro generation or distributed generation	Large-scale generation or central generation
Transmission losses	Lower losses	Higher losses (in India, around 26% is lost in aggregate technical and commercial
Changes occurring in grid	The transmission load is reduced, so the need for grid upgrades is lower	The transmitted power is increased, so there is a need for grid upgrades
Waste heat as byproduct	Can be used for heating purposes, commonly known as micro combined heat and power	Used in privately owned industrial combined heat and power (CHP) installations
Reliability and management	Extremely reliable and managed by an individual, a group of customers or small industries	Managed by power company and government
Future development	Focused on the greenness of energy	Focused on energy crisis
Ability to meet needs	Renewable energy resources are able to produce lesser than nameplate capacity	Fuel-based or hydro-power generation is fully dispatchable

2.2 HISTORY OF MICROGRIDS

The history of microgrids started in 1955 at the Whitling refinery, which is located in Indiana, where the first modern industrial microgrid with a 64 MW facility was constructed in the United States per Pike Research. But it was not where the first microgrid was spotted; rather, the concept of a microgrid started with the beginning of industry. In the year 1882, no standards for a generation and distribution system had been set for electricity, so Thomas Alva Edison designed his Pearl Street Station as he went along [8]. The station was situated at 255–257 Pearl Street in Manhattan, and the station was powered by coal. Surprisingly, the design of Edison's Pearl Street station met all of today's criteria for a microgrid system. The coal-fired steam engines in the station drove six jumbo generators, and each generator was able to produce 1100 KW DC. On September 4, 1882, the Pearl Street Station began operating, and it provided electricity to 508 customers who were present in New York, and a total of 10,164 lamps were provided with electricity. The byproduct of the plant was found to be steam, so Edison also made the use of this steam by providing it to local manufacturers, and the heat produced from the steam was provided to nearby buildings [8]. It can be seen that the electric system created by Edison was a small localized generation unit with a distribution network that was limited to a small area, a perfect replica of today's microgrid definition and structure.

Due to the restrictions of the DC transmission system, Edison's system was able to serve only a few blocks in any direction, and his system also included storage batteries. This concept gave rise to the idea of microgrids with energy storage and interestingly almost all today's research deals with this topic. Today we use an alternating current (AC) grid, which makes the biggest difference observed from the electric system built by Edison because he used a direct current (DC) grid. DC grids were able to work only in densely populated areas like Lower Manhattan, as shown in Figure 2.2, and the voltage levels were low. It was nearly impossible to implement a DC grid on a wider scale.

So, for long-distance transmission, the AC grid built by Nikola Tesla proved the best solution in 1893 at the Chicago World's Fair, which shaped history as we know

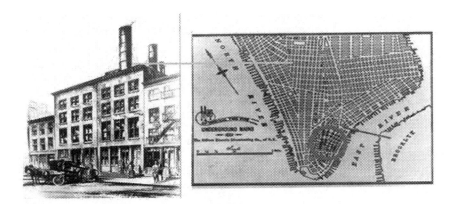

FIGURE 2.2 Pearl Street Station and Lower Manhattan, which was lit up by Edison.

Microgrid Design Evolution and Architecture

FIGURE 2.3 Edison's DC grid connection for Pearl Street station.

it today. Today's AC grids are the outcome of that history. But the power demand is increasing day by day, and customers expect an uninterrupted power supply, which inspired the idea of microgrids, and the outcome of these microgrids is DC. Due to developments in power electronics, today it is possible to connect DC and AC by using rectifiers and inverters. Today, microgrids can integrate with renewable energy plants and also with distributed generation plants where the outcome is combined heat and power, which in short is termed CHP—this concept of CHP also came from the Edison's Pearl Street station, as shown in Figure 2.3 [8].

2.3 NEED FOR MICROGRIDS

Today there exist some places with lack of electricity services, mainly rural areas and semi-urban areas of underdeveloped and developing countries. According to the International Energy Agency, nearly 940 million people (13% of the total population) across the world and 240 million people in India do not have access to electricity in 2017. The extension of the central grid in many of these places is the traditional approach, which is technically and financially inefficient due to the existing insufficient energy service along with reduced grid reliability, extended building times and construction challenges in remote areas. Many of the challenges faced by traditional lighting or electrification strategies can be overcome by adequately financed and operated microgrids based on renewable and appropriate resources. The world is expecting universal access to electricity by 2030, but progress towards it is quite slow. To reach the goal of universal accessibility, microgrids play a vital role. Governments, NGOs and other private developers throughout the world are focusing on microgrids to electrify communities that are likely to be served in the near or medium term by extensions of traditional centralized grid systems. As a result, the number of microgrids being developed is increasing rapidly. Although the power rate of microgrids is limited to a few mega-volt ampere (MVA), it is relative to its application area and grid type. To meet higher power demands, there is need of a large number of microgrids, and these are interconnected with increased stability and necessary control technologies; this interconnection is referred to as power parks. The involvement of renewable sources makes the microgrid eco-friendly by decreasing environmental emissions. The provision of energy security for local customers or communities is also provided and operated without the involvement of a central grid. The reliability, sustainability and economics are the most important goals for the electrical society, and these three goals can be fulfilled with microgrid technology [9–11]. When compared to conventional grid applications (where only 33.33% of fossil fuel consumed is converted to electricity and the remainder is dissipated as heat energy), microgrid technology can interact with consumers on its distribution side, and therefore it can manage the power demand and supply. Not all the power generated in the main grid is utilized completely by consumers because about 5–7% of power is lost during transmission in the transmission lines. Coming to a microgrid, all the power generated stays at the distribution level, and 20% of microgrid generation capacity helps to meet the 5% peak demand in the utility grid. One of the most important features of microgrids is islanded operation or independent operation during events of widespread failure or power fluctuations (intentional or unintentional) or even for instants of cost optimization. The microgrid enables a black start facility, which might be required during disasters or calamities.

Microgrid capacity ranges from one to a few hundred kW depending on the number of customers connected and type of loads attached. The services provided by the microgrid include residential lighting, fans, refrigerators, TVs and so on. Depending on the favorable conditions of the place and with respect to the capital cost, the type of generation resource employed varies. Generation resources could be solar PVs, windmills, biomass gasification, hybrid technologies such as solar PV and diesel or wind diesel. All are familiar with diesel-based microgrids due to the widespread availability of diesel and low upfront capital cost when compared to other microgrids. It is also well known that diesel generators (Figure 2.8) may last for 40 years or more, because the trend of renewable energy sources for backup generation during power outage is becoming successful. Microgrids with renewable energy source–based generation are drawing more interest. Micro-hydro-based microgrids are generally situated near river areas and are much more similar to run-of-the-river–type schemes where a pipe is used to divert the direction of the water from a river or a stream and this water is allowed to flow through a turbine to generate electricity. Biomass gasification systems use biomass that has undergone incomplete combustion to produce syngas, and this gas is burned in an engine to run a generator. A biomass-based microgrid is suitable for areas where an adequate biomass supply is available. Solar PV–based microgrids are globally known and in widespread use due to the reduction in the global market price of PV cells. With modern technologies, the price of solar cells will be reduced even more. PV cells use solar energy and convert it into electric energy, which is stored in batteries and used as backup. Wind-based microgrids are no less popular than solar-based microgrids. Windmills convert wind energy into electric energy, and it is stored in batteries. A battery storage system is required in both solar- and wind-based microgrids for uninterrupted power supply and for backup plans during power outages. There are some prospective features of any electric grid, which include the involvement of independent microgrids with ease of connecting and disconnecting [12, 13].

Microgrid Design Evolution and Architecture

The features also include the following:

1. The harnessing of power and heat simultaneously as in combined heat and power technology.
2. Minimization of transmission losses.
3. Response to energy demand.
4. Integration of renewable energy resources.
5. Power quality enhancement and improvement.
6. Future prediction of generation and load.
7. Resilience to grid failures.
8. Increase in number of active consumers in power production and stakeholders in transactions of electric energy.
9. Improvement in energy storage devices and so on.

The mentioned features can be fulfilled by the development of microgrids into the existing utility grid [12].

2.4 DEFINITIONS OF MICROGRIDS

Microgrids can be understood by different definitions given by different departments or research, but the ultimate goal is to construct a sustainable and efficient microgrid. Two definitions of microgrids are given in the following.

Per the United States Department of Energy's microgrid exchange group, a microgrid is defined as a group of interconnected loads and distributed energy resources within clearly defined electrical boundaries that acts as a single controllable entity with respect to the grid. A microgrid can connect and disconnect from the grid to enable it to operate in both connected and islanded mode.

The European Union research project has a definition of microgrid in its terms, stated as follows: "A microgrid comprises low voltage distribution systems with distributed energy resources, storage devices, energy storage systems and flexible loads and such systems can operate either connected or disconnected from the main grid."

In a short, a microgrid can also be defined as a modern power distribution system where local sustainable energy or power resources are designed with the help of various smart grid initiatives. In all the definitions, the term "distributed energy resources" is used, and it refers to power generation situated near the consumption site or nearby loads. Main power generation in a centralized grid can be referred to as central generation, and power generation situated near the distribution network as a part of a microgrid is referred to as distributed generation.

2.5 STRUCTURE OF MICROGRIDS

A microgrid is an interconnection of different renewable distributed generators, non-renewable distributed generators with devices capable of storing energy and various types of microgrid loads. The structure of a microgrid also includes the interconnection of other microgrids, control and stability systems and communication systems between the consumer and owner. The interconnection between the main grid and

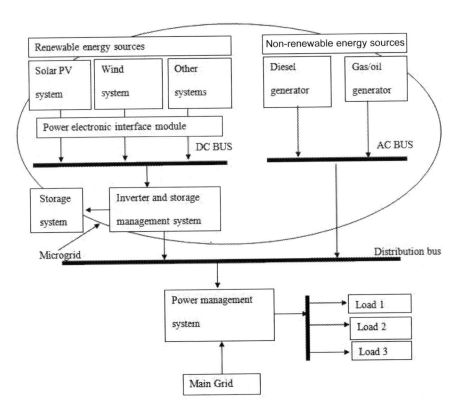

FIGURE 2.4 General structure of a microgrid.

the microgrid is known as a point of common coupling [1]. The general structure of a microgrid is shown in Figure 2.4, where both renewable and nonrenewable energy sources are distributed energy resources.

The construction of a microgrid involves the components of distributed generation, energy storage devices, converter technologies, power management systems and control technologies for stability and efficient operation of the microgrid.

2.5.1 Distributed Generation

"Distributed generation" refers to generation of power from nearby consumer regions. It reduces or eliminates the generation cost: distribution cost along with transmission cost when compared with existing main generation units. It also increases the efficiency by the removal of elements that are the reason for the complex and interdependent nature. In most situations, distributed generation may provide lower generation costs with more reliability and enhanced security, which is not achieved by the existing generators. Distributed generation uses small-scale technologies for production of electricity near the end users of power. For an increase in potential benefits, distributed generation technologies include modular generators and sometimes

Microgrid Design Evolution and Architecture

renewable energy resources. When compared with traditional power generators, distributed generation can offer lower-cost electricity and higher power reliability and security with fewer environmental consequences, as shown in Figure 2.5.

Distributed generation technologies include reciprocating piston engines, gas turbines, fuel cells, wind turbines, PV systems and so on. A point of common coupling is an important aspect in interconnecting distributed generation with the main utility grid. Distributed generation takes place on two levels, the local level and the end point level, as shown in Figure 2.6. Local-level power generation plants need to be located specifically, and accordingly, renewable resources must be used per the locality. These include solar power generation, small hydro thermal plants and so on. Local-level plants are less centralized and tend to be smaller, which makes them cost and energy efficient with more reliability. The end point level is very similar to the local level, with many common effects, since the individual energy consumer at the end point level applies many of the same technologies stated for the local level. Mostly the modular internal combustion engine is a frequently employed distributed generation by endpoint users. Distributed generation can also operate as an isolated island of electric energy production and serve as a small-scale contributor to the power grid. In the traditional electric power generation approach, a few large-scale generating stations are installed far from load centers, but in contrast, distributed generation systems use numerous but small-capacity plants and can provide onsite power with less reliance on the distribution network and transmission grid. A general layout of distributed generation interconnected with a utility grid is shown in Figure 2.7.

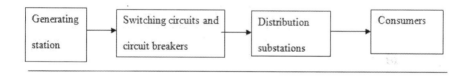

FIGURE 2.5 Schematic illustration of conventional electricity grid with power generation and distribution.

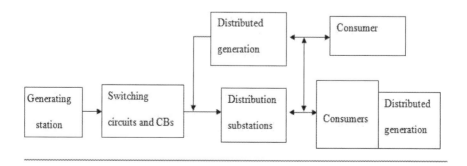

FIGURE 2.6 Schematic illustration of power supply by main grid and distributed generation.

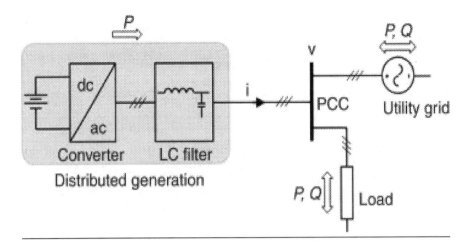

FIGURE 2.7 Interconnection diagram of distributed generation, utility grid and load by PCC.

The typical uses of distributed generation include domestic purposes such as micro generation of electricity and heat, commercial purposes (building related), greenhouses, industrial purposes, district heating and grid power. Some of main generation technologies which can be used as distributed generation are discussed in the following.

2.5.1.1 Stand-Alone Reciprocating Engine Generators

The concept of distributed generation is neither so old nor too new, because with the beginning of electric grid establishment, there were some remote communities, or off-grid communities. For those communities, stand-alone reciprocating engine generators found application as power plants for these communities [14, 15]. Perhaps these units may be the first crude incarnation of distributed generation. These engines run with fossil fuels, which make them relatively inefficient, and these units are also responsible for significant local pollution and need to be replaced by modern technologies or renewable distributed generation.

2.5.1.2 Turbine Generators

Turbine generators that are run through gas or gas turbines have been much more familiar in the aerospace industry. In the context of application as distributed generation, these are highly efficient, less expensive and possess mass-produced technology. Turbine generators can achieve a thermal efficiency up to 60% by utilizing the combined generation of electricity from the work of turbine shaft and electricity from the exhausted waste (through a steam cycle). If the waste heat from domestic and commercial heating is utilized to generate electricity, then the efficiency can rise up to 90%. These units can be operated as a medium-sized cogeneration plants. Micro gas turbines (Figure 2.9) can be ideal for small businesses or households, which offers new opportunities for peak shaving and cogeneration at a more localized level.

Microgrid Design Evolution and Architecture 29

FIGURE 2.8 Diesel generator.

FIGURE 2.9 Microturbine.

2.5.1.3 Solar PV Cells

Solar PV (Figure 2.10) cells convert solar light energy into electrical energy directly with the help of solar PV modules or solar PV arrays. Solar PV panels are expensive due to their sophisticated manufacturing technology. But with the increase in market demand and modern manufacturing technologies, PV cells are going to become more economically appealing options in the near future. An efficient solar power plant can be seen as a green electricity utility for many consumers, but the capacity factor of solar PV cells makes solar power generation a potentially troublesome generation technology. Use of energy storage devices in these units makes it a more practical option.

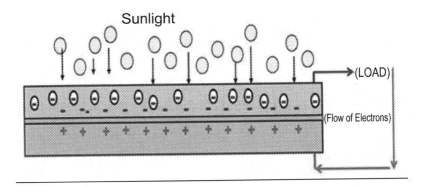

FIGURE 2.10 Solar PV cell.

FIGURE 2.11 Wind turbine.

2.5.1.4 Wind Turbines

Apart from solar PV cells, there is an indirect approach to utilizing solar energy, and that approach is converting wind energy, which is an outcome of solar energy, into electrical energy. The use of wind turbines (Figure 2.11) and energy storage devices is relatively expensive, but the outcome is much more beneficial. The utilization of wind power within the infrastructure of distributed generation is limited with respect to geographical regions (with regular and high-quality wind conditions) or to off-grid communities.

2.5.1.5 Fuel Cells

Compared to renewable technologies, fuel cell technologies are highly efficient. Here the chemical energy is converted into electrical energy with heat as a byproduct.

Microgrid Design Evolution and Architecture

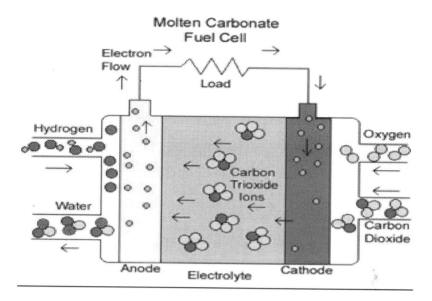

FIGURE 2.12 Fuel cell technology.

The fuel cells generate electricity by means of chemical reactions taking place by transferring charged species across a membrane which separates the fuel and oxidant streams. The fuel used here is hydrogen or methane. The demerit in this technology is the risk factor and the pollutants formed from the side reactions. If those pollutants are avoided, then electric efficiency can be as much as 55% for fuel cells. Recently fuel cells (Figure 2.12) are being explored for use in vehicles and also as a stationary generator. Their use is limited due to the high capital cost.

2.5.2 Energy Storage Technology

For successful operation of a microgrid, energy storage technology plays a vital role. The electrical energy available from renewable sources is variable and depends on environmental conditions. So for a smooth supply and in order to legitimize renewable energy sources as a reliable contributor to the main grid, energy storage is important. Continuity in the power supply is possible by balancing power generation with power demand, and without an energy storage process, it is quite difficult [16–19]. The energy storage components present in a microgrid are required to meet the following requirements:

1. The foremost requirement for an energy storage device is to balance power generation at the generation side with power demand on the load side because most distributed generation used in microgrids is intermittent and prone to transient disturbances.

2. Energy storage devices must be able to store the maximum energy demands during off-peak hours and be able to supply all the consumers or electric loads whenever required.
3. To meet unpredicted and sudden load demands, the energy storage device must be able to eliminate the loaded parts present in the microgrid.
4. It should be able to accommodate the minute-hour peaks that are present in the daily load demand curve.
5. Whenever there is a switching of a microgrid from grid-connected mode to islanded mode or vice versa, the energy storage device has to provide smooth transient conditions.
6. It should be able to ensure sustainability and reliability in terms of constant voltage and frequency operation when the distributed generation is renewable.

The classification of electrical energy storage technologies is mainly based on its construction, and they are:

1. **Electrical technologies**: The electrical storage system includes the super capacitor system and super-conductive magnetic coil. A suitable storage system in microgrids is a super capacitor, where the electrical energy is stored in the form of a static electric field in the electrolyte between the electrodes and ions. Super capacitors have very high cycle life duration with high power density but with much lower density of energy. That's why they are used in microgrids where a short-term and high-power storage system is required. They can also be used in hybrid storage techniques with batteries so that their lifetime will be increased [21].
2. **Mechanical technologies**: The mechanical storage system includes the pumped hydro-power system, compressed air energy storage system and flywheels. Pumped hydro-energy storage is considered the major storage technology across the world with more than 127 GW installed power. It is used as medium-term storage system due to the range of its energy-to-power ratio from 2 to 8 hours. It consists of two interconnected water reservoirs which are located at different altitudes, for example, a mountain lake and a valley lake, and is connected by using penstocks. This storage technology is used when the distributed generation used is a small hydro-power plant. It is designed for collecting water during the wet seasons so that hydro-power can be produced continuously throughout the year. There is an alternative to hydro-storage systems called a compressed air energy storage system (CAES). To date, only two CAES plants are in operation, one in McIntosch, US, with capacity 110 MW and the other in Huntrof, Germany, with capacity 320 MW. For stabilization functions in weak grids or in distributed generation where renewable sources are used, the mechanical storage system used is a flywheel storage system. The rotating kinetic energy of the flywheel is stored and used during high power demand for short durations with charging discharging cycles [21].

3. **Chemical technologies**: A chemical storage system includes batteries, hydrogen storage and natural gas storage system. The batteries can be classified as lead acid, lithium ion, high-temperature and flow batteries. The major development in battery technology is due to the major application of battery storage systems in microgrids involving renewable energy sources as their distributed generation. Hydrogen storage systems are also chemical energy storage systems, which are expected to be more important for power systems involving high fractions of renewable energy of about 80–100% [21].

2.5.3 Microgrid Loads

Although microgrids are constructed for backup purposes and for an uninterrupted power supply, the loads connected to the microgrid should be analyzed properly, and the type of load should be chosen for proper reliability. Various kinds of loads can be connected to a microgrid system, which plays an important role in the operation, stabilization and control of the microgrid. The power system contains different kinds of loads, which are classified as static and dynamic loads. The loads that can be supplied include household and industrial loads, but these loads are critical or sensitive and demand reliability at a high level. The operation of a microgrid requires several considerations with respect to loads. Some of the considerations are priority to critical loads, power quality improvement for specific loads and also enhancement of reliability for pre-specified load categories. In addition to this, distributed generation or local generation also prevents some unexpected disturbances with the help of fast and accurate protection systems [20]. For the prediction of operating strategy in a microgrid arrangement, the following considerations are to be seen before and so that load classification is possible:

1. The load or source operation strategy included in the microgrid is required to match the net active and reactive power in grid-connected mode.
2. The load/source operation strategy included in the microgrid is required to meet the stabilization of the voltage and frequency in islanded mode.
3. The type of load should be a category of power quality improvement type.
4. With respect to the distributed energy resource ratings, there should be a reduction of maximum load for better enhancement.

2.6 OPERATION OF MICROGRIDS

The operation of microgrid involves two operations:

1. **Grid-connected or grid-tied mode:** Per the name of the operation mode, the utility grid or main grid is connected to the microgrid, and the microgrid can demand energy from the main grid, or it can supply energy to the main grid. During this transfer of energy, if there is any disruption occasionally or intentionally, then the microgrid is switched to islanded mode. Since here the main grid is connected to the microgrid, it is termed grid-connected

mode [25]. There lies a point of common coupling where the microgrid and utility grid are connected, as shown in Figure 2.13.

2. **Islanded mode:** Per the name of the operation mode, the microgrid gets disconnected from the main grid either naturally or intentionally, and the work of the microgrid is autonomous. The ultimate aim is to supply the amount of energy demanded by local needs during the absence of the main grid and not to rely on the main grid, as shown in Figure 2.14. Once the disruption or the disturbance in the main grid is cleared, then the microgrid will get connected to the main grid and thus return to grid-connected mode [25].

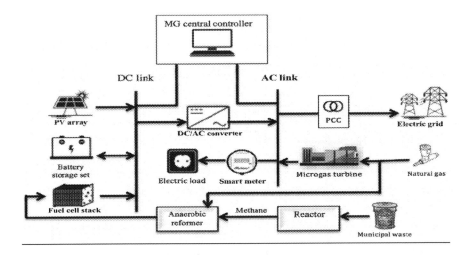

FIGURE 2.13 The point of common coupling connecting microgrid and utility grid.

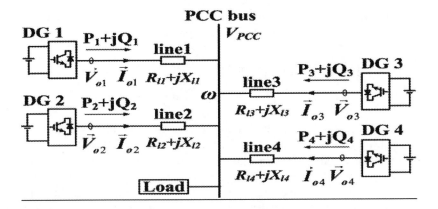

FIGURE 2.14 Operation of distributed generation in islanded mode of operation.

The switching of the microgrid from grid-connected to islanded mode or vice versa needs a proper monitoring system and a control strategy, which are discussed in further sections.

2.7 MONITORING OF MICROGRIDS

For the suitable operation of a microgrid, a monitoring system is required to match the following tasks and requirements:

1. It is required that the monitoring system use a distributed structure so that it is possible to monitor individual generation units simultaneously, and reliability, along with microgrid system's operational efficiency, will also improve. By using a distributed structure, it is also possible to monitor the operating parameters comprehensively.
2. The monitoring of the microgrid system must be able to support different varieties of protocols related to the communication system. The support of communication protocols is necessary to ensure communication between the user and terminal device located at the distributed side or distributed generation, which includes the receiving signal parameters (either in analog or digital form) and handling them in different formats.
3. The configuration of the microgrid system should be defined, and the configuration of different subsystems and devices present in the main system should be defined. An online modification facility must be present where the configuration of all types of tasks to be performed is defined.
4. The monitoring of the microgrid also includes the monitoring of the main grid operating system, individual subsystem operation and every device involved in the microgrid system. It enables the remote controlling of the microgrid and monitors the switching modes of the microgrid.
5. Microgrid system state prediction is required for the promotion of operation of the microgrid system and also for its control. So, to make accurate predictions, one must analyze the previous state of the system, which is possible from the stored operation data from time to time. This requires the continuous monitoring of the microgrid system [21–24].

Hence, a monitoring system is required to record and store the long-term operating data of the microgrid. A schematic block diagram of a monitoring system is shown in Figure 2.15, including the use of peripheral component interconnect extensions for instrumentation (PXI), which are PC-based technologies that help in storing a high number of channels with high performance and high speed. Apart from state prediction, the stored operation data can also be used to analyze the causes of accidents, for analysis of statistics related to the operation and economy of microgrid and for future modifications or developments. The equipment present in the subsystem is considered a monitoring device, and distributed generation is considered subunit monitoring [25]. The core of the monitoring system is referred to as the inverter because the grid-tied inverter is known as the heart of all distributed generation units connected, and it stores almost all operating and running data. For detection

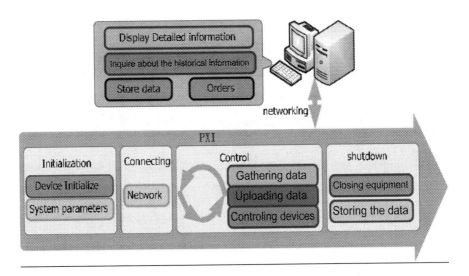

FIGURE 2.15 The use of peripheral component interconnect extensions for instrumentation in monitoring of a microgrid.

of the status of the switch that transits the mode of operation of the microgrid and for remote monitoring, the data recorded from the communication between various devices and distributed generation ports is used. The control system of the microgrid sets different control processes and specific programs for continuous monitoring and operation of the microgrid.

2.8 STABILITY, CONTROL, AND STRATEGIES FOR MICROGRIDS

After discussing the monitoring of microgrids, the issues to be focused on most importantly in the context of microgrids are stability issues. Stability issues are more common in microgrids compared to main utility grids because microgrids have much lower energy and power ratings than main grids. Microgrids can be of two types based on the nature of their output power, AC and DC microgrids [26]. The concept of stability issues in AC microgrids and their analysis are much more similar to that of main grid. Active and reactive power controls can regulate the voltage and frequency of the system. The stability can be controlled by torque and speed control of the shaft of the machine only if the distributed generation used is a conventional generator (standalone generator) and is connected directly without a power electronic interface module. In DC microgrids, there seem to be fewer or no stability issues because there are no interactions of reactive power in DC grids. The regulation of system frequency is important in DC microgrids. Apart from the stability issue, the major issue faced by any microgrid is power quality, and this issue is also raised during the interconnection of distributed generations. Based on the regulation of voltage and frequency, stability is classified into two types, voltage stability and frequency stability. The changes in system frequency occur due to transients and small signal disturbances, for example, during switching a microgrid from one mode to another, and the changes are to be eliminated, which is termed frequency stability. The stability of a microgrid is generally classified as one of two types. Small signal stability methods, which are analyzed from the basis of closed-loop controllers, can be used to handle the problems of continuous switching loads and manage the demand of power by distributed generations or micro sources. Transient disturbances can be seen during a fault occurring in any one of the subsequent microgrids that are in islanded mode, and this affects the entire microgrid system in the name of transient stability. The voltage stability problem in microgrids can be caused due to transient sources like tap changers or due to changes in load (or load dynamics) [30–33]. Also, by limited reactive power, if there is a mismatch in the specified voltage levels of a microgrid and utility grid, then voltage stability or voltage regulation is required. By using well-developed and suited control strategies with additional closed-loop controllers and observers, small signal stability can be enhanced. By using proper or advanced storage devices and suitable protection devices for the microgrid, transient stability can be improved. To improve the voltage stability of the microgrid, voltage-regulating devices and reactive power compensators can be employed. In addition, current limiters and load controllers can also be employed to ensure of microgrid stability.

The strategy of control systems used in microgrids requires the selection or optimization of an operating point for the microgrid either manually or autonomously. For insertion of new distributed generations or micro sources or for removal of old distributed generations from the microgrid without any moderation in the existing components of the system, the operation of a microgrid with proper control strategy is needed. There is a requirement for controlled operation of the microgrid for independent control of active and reactive power, for correction in voltage sags and for meeting the load dynamics that are involved in a utility grid [6,7]. To control the operation of the switch that connects or isolates the microgrid from the utility grid per the demand, control technology for a microgrid is a must because it allows smooth switching during demand or intentional islanding and immediate switching during faults or unintentional islanding. The current developments in the microgrid area are mostly related to different control strategies that are to be employed in microgrids for efficient operation. Different control strategies include independent control of individual distributed generation sources, improvement strategies in the central control system and agent-based observing system strategies. Independent control of individual distributed generation sources mainly refers to the governing control of traditional generators by itself, which improves adjustment to the existing grid system with variable conditions. Improvement in the central control system deals with the increase in infrastructure related to communication strategies and easy integration of the increased infrastructure into the main system. An agent-based observing system implements remote or local control of the microgrid at various levels. Due to this method, the robust nature of centralized and distributed control systems can be exploited.

The control systems used in microgrid systems can be classified into three categories depending on the controllers used: localized control systems, centralized

1. **Localized control system:** For better performance and efficiency of a microgrid, the localized control system plays a vital role in controlling local distributed generations or micro sources. The controllers involved are referred to as local controllers (LCs), and this operation of this control system can be carried out in the absence of any software or any system, as in Figure 2.16. These controllers have interaction with other controllers, so this allows them to work even in the islanded mode of a microgrid. Sometimes this localized control system is also called a decentralized control system because here all the controllers have independent control operation, as in Figure 2.17. It can call an operation or deny it from the point of distribution with respect to the situation, which is a fast solution for the entire system. Conversely, a complete decentralized or localized control system can have more problems because all the control operation is at the distributed level, which may lead the manager to lose control over all the distributed generations and can affect the complete microgrid system. To overcome this situation, a well-organized control system is required to control or govern all the local controllers, which gave rise to the centralized control system [27–29].

2. **Centralized control system:** As the name suggests, the control system is centralized and the controllers involved here are referred to as microgrid

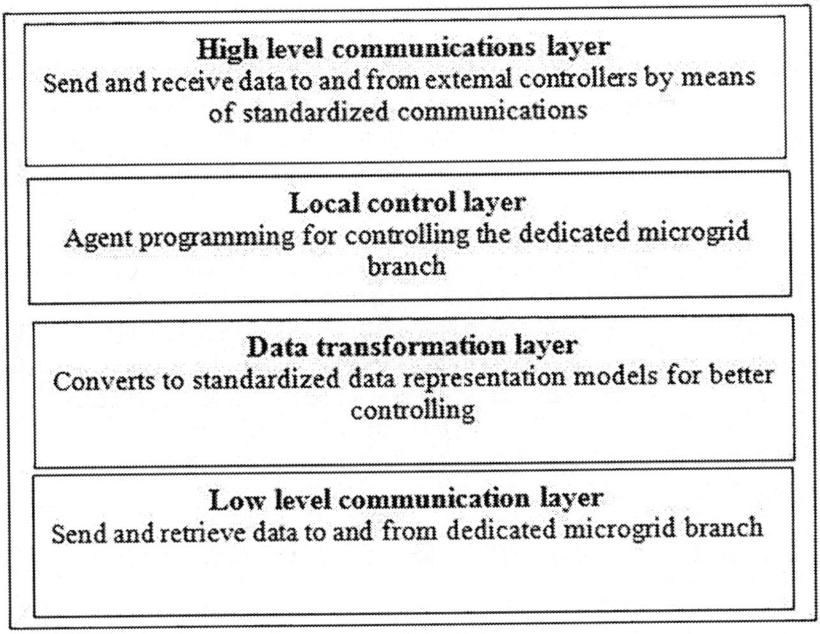

High level communications layer
Send and receive data to and from external controllers by means of standardized communications

Local control layer
Agent programming for controlling the dedicated microgrid branch

Data transformation layer
Converts to standardized data representation models for better controlling

Low level communication layer
Send and retrieve data to and from dedicated microgrid branch

FIGURE 2.16 Different layers of local control operation.

Microgrid Design Evolution and Architecture

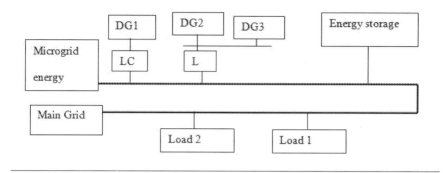

FIGURE 2.17 Location of local controllers in microgrid.

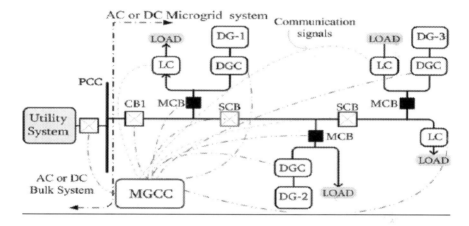

FIGURE 2.18 Centralized control system architecture.

central controllers (MGCCs). To integrate this system into the main system, the system has to be hierarchic in nature, as shown in Figure 2.18. All the local controllers and distributed generations along with loads are interlinked with the MGCCs and provide control over their operation. A centralized control system can achieve comprehension from a specified central point that usually depends on the type of network, and it could be a controller, a server or a switch. The installation of a centralized control system helps the operator control the entire system to meet the necessary power demands with ease. However, there is a solo control device or single MGCC at a centralized control system that stores all the data, measures it and processes it. This operation of a unique controller device may cause several communication problems in centralized control systems and could cause severe faults, which may lead to shutdown of the entire system [34–36].

3. **Distributed control system:** In a distributed control system, the MGCC is absent, but interconnection between the local controllers is present for a

communication strategy between the LCs. In this control system, LCs can communicate directly with one another and can coordinate or negotiate for optimal control over the microgrid. If there is any single point of failure, then distributed control is the best backup for it. The three control systems at a glance are shown in Figure 2.19. However, the construction of distributed control systems deals with numerous complexities in terms of analytical performances, and communication lags between LCs. The speed of this system depends on the communication speed and operation speed of LCs [34].

4. **Multi-agent-based control system:** There exists another concept of control system which involves the properties of both centralized and localized control systems and is known as a multi-agent-based control system (MAS). It is similar to a distributed control system, but it involves the operation of agents which operate depending on the data achieved from LCs and send the control to the respective controllers. As shown in Figure 2.20, there

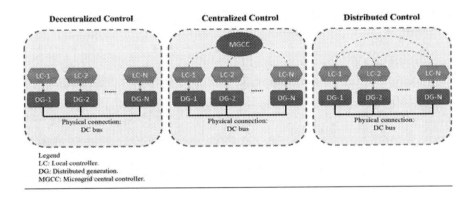

FIGURE 2.19 Difference between three control systems.

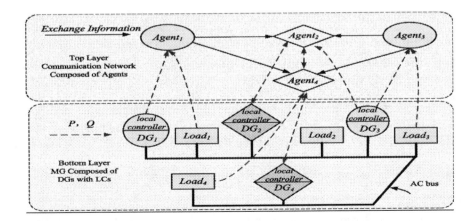

FIGURE 2.20 Multi-agent-based control system.

is interaction between controllers, with each helping to improve the infrastructure of the communication strategy in a wide area where the voltage stability is taken care of by MAS [35].

Numerous research works and models have proposed control strategies for microgrids to enhance their performance and efficiency.

2.8.1 Microgrid Control Techniques

In a microgrid, distributed generations are connected with a bus bar through the power electronic interface module, which is different from central generators that are connected directly to the main grid without any interface. The necessity of energy storage devices in the microgrid was discussed earlier. That's why conventional methods are not employed in microgrids as in main grids. There exist several control technologies or techniques to be employed in the control systems of microgrids. The techniques can be classified as master and slave control, current and power flow control and peer-to-peer control or droop control [37]. In master and slave control, the circuit operating as master regulates the voltage values and values of frequencies, and the circuit operating as slave controls the current sources present in the system. The current and power flow control method controls the power distribution among the devices and current distribution by employing certain control signals. The peer-to-peer control method is completely based on the method of external characteristics of declining. Peer-to-peer control methods manage the conditions of a single point of failure of distributed generation connected in the microgrid. If single-agent failure occurs in the system, then the information and communication won't also be interrupted in to peer-to-peer connection model. Since it is based on droop characteristics, it can also be termed the droop control method. The droop control method is combined with either the master and slave control method or current and power flow control method to improve their control availability because the system includes power converters which act as non-ideal voltage sources [38–41]. With the help of a specific control algorithm, the system frequency and voltage can be adjusted autonomously without the involvement of communication. Drooping characteristics are seen in f/P (frequency versus active power) control and V/Q (voltage versus reactive power) control where the reference P and Q of distributed generations are produced by measuring the frequency of the microgrid system and the magnitude of output voltage of respective distributed generations. There is an alternative control method, P/f and Q/V, where the reference frequency and magnitude of voltage are produced by measuring the respective P and Q of distributed generations.

2.9 INTERFACE MODULES OF MICROGRIDS

The architecture of a microgrid is not complete without studying the importance of interfacing modules used to connect the distributed generation and bus bar of existing grids. These modules used in microgrids, as shown in Figure 2.21, are made of power electronic converters which allow the interconnection of power equipment and components of the common system. Most distributed generations are renewable

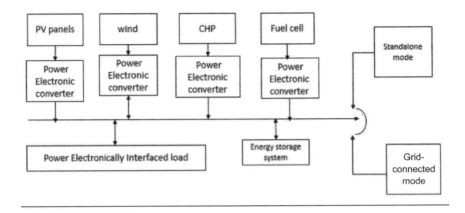

FIGURE 2.21 Microgrid with power electronic interface module.

resources, and they require specialized or advanced power electronic converters to convert the energy generated to a suitable energy form utilized by the consumer or supplied to the utility grid. The use of advanced power electronic converters as interface modules allows the system to meet necessary power demand at better efficiency and at low cost. Since in microgrids different varieties of sources are employed, the use of a power electronic interface will bring stability to the system [41–42].

Distributed energy resources where distributed generations and energy storage devices are used to meet the required power demand involve the conversion of DC energy to AC energy and vice versa. Grid-tied inverters are used to convert the energy generated in the microgrid to a suitable AC form which is compatible with the utility grid. The power electronic interfaces also control the voltage and frequency stability of the microgrid with the grid. In addition, the interface modules have several functions other than conversion of power, such as maintaining power quality, power conditioning of active and reactive power, controlling load dynamics, monitoring and controlling services, protecting output filters and interfaces and so on. The interdisciplinary nature of power electronics helps in changing the voltage characteristics, current characteristics or any characteristics of power into suitable quantities per the power application. Among the power electronic devices used as interfaces, bidirectional converters are assumed to be widely used components in microgrids due to their ability to control power flow. Also, bidirectional converters are able to handle microgrid-generated power in a stable manner either during low or no load operation mode or overload operation mode.

The interface takes control of the microgrid when it is disconnected from or reconnected to the main grid. In general, the microgrid operates in grid-connected mode, and if any disturbances occur intentionally or unintentionally, then the microgrid is disconnected from the main grid at the point of common coupling. Until the disturbance is cleared, the microgrid operates in islanded mode [42]. Typically, the rated peak power capability of a microgrid is limited to 10 MVA, so microgrids are generally connected at distribution levels with limited handling capacity of energy due to usage of renewable energy resources and their wastage of heat. Hence the relay used

Microgrid Design Evolution and Architecture

for the interconnection of the microgrid that interfaces the microgrid with public utilities plays a vital role, and it determines the success rate of microgrid management.

2.10 TYPES OF MICROGRIDS

The types of microgrids based on the locality and number of communities are discussed in the following:

1. **Customer microgrids:** The general structure which is discussed in this chapter is a perfect example of a customer microgrid. These are often called true microgrids. Usually customer microgrids are self-governing and situated downstream of a single PCC. One can imagine this type of microgrid easily because the regulations and structure involved here can neatly fit into our present technology. The restrictions on the nature of the microgrid are comparatively neglected or loose.
2. **Utility or community microgrids:** Community microgrids are a small section of regulated grids. They are also known as milligrids. Due to the presence of traditional incorporate infrastructure, community grids are different from a statutory and business model view fundamentally, but technically these are no different from microgrids. The inference here is that the regulation of the utility is much more significant and the microgrid has to comply with utility codes that exist, or, if possible, accommodation has to be made to that code.
3. **Virtual microgrids:** This is the type of microgrid which covers distributed energy resources at numerous sites, but all are coordinated with each other in such a manner that the grid sees them as a single controlled entity. To date, very few demonstrations of virtual grids exist and more development and control techniques are required. This type of system must be able to operate in a controlled islanded mode of operation or in coordinated multiple islands.
4. **Remote power systems:** The concepts and structure of remote power systems are fairly close to the concepts that are found in microgrid technology; that's why remote power systems are commonly defined as microgrids. These can also be termed remote microgrids, and these are not operated in grid-connected mode. These power systems work in islanded mode only. These are located in highly dispersed consumption regions.

Per A&S Utility distribution microgrids (as of 2012), microgrid categories can be of three different types other than the ones mentioned previously, as follows:

1. **Campus environment or institutional microgrids:** The main focus of institutional microgrids is the accumulation of existing onsite generation with more than one load located near the same geography so that the manager can easily control and manage the microgrid.
2. **Military base microgrids:** Military facilities demand a continuous and uninterrupted power supply for physical and cyber security. So, to fulfill this function without depending on a utility grid, microgrids deployed near military base camps are referred to as military base microgrids.

3. **Commercial and industrial microgrids:** There are many industrial and manufacturing processes where an interruption of the power supply for a short duration can cause high economic losses to the industries. So industries build microgrids per their demand with necessary standards to avoid interruption in the power supply, and those microgrids are called commercial or industrial microgrids [43–46].

2.11 PROS AND CONS OF MICROGRIDS

Microgrids also have two sides, one side with numerous benefits or advantages and the other with disadvantages. Some of the pros of microgrids as identified by Lawrence Berkeley National Laboratory (LBNL) are discussed in the following:

1. **Autonomy:** A microgrid allows the autonomous operation of generation devices, storage and loads seamlessly, and with the help of recent advances in technology, microgrids balance out voltage and frequency stability issues.
2. **Stability:** Irrespective of the main grid in the on or off state, the entire micro-system operates in a stable manner due to the utilization of control techniques that are based on droop characteristics of frequency and voltage levels at the terminal of each device.
3. **Compatibility:** Microgrids operate as functional units to existing grids, which give assistance in building the existing system and thus helping it to maximize with standard utility assets.
4. **Flexibility:** Microgrids are flexible, with easier concepts of electricity generation, and can use hybrid technologies by combining more than one type of distributed generation.
5. **Scalability:** Microgrids provide the facility of distributed generations, energy storage devices, and load devices in a parallel and modular manner, which makes it easy to scale up with higher-power generation and/or consumption levels.
6. **Efficiency:** Energy management goals—including economic and environmental—can be optimized in a systematic fashion. Blackouts and power interruptions are minimized
7. **Economics:** The improved reliability of the microgrid significantly reduces the cost experienced by consumers due to power outages and poor power quality. It also leads to active participation of consumers, which builds up economic efficiency.
8. **Peer-to-Peer Model:** Microgrids represent a new paradigm—a true peer-to-peer energy delivery model that does not dictate size, scale, peer numbers or growth rates.

Some of the cons of microgrids are discussed in the following:

1. The controlling and consideration of three main parameters: frequency, voltage and power quality, such that they meet with accepted standards, but at the same time, the power and energy balance are to be maintained.

2. The requirement of more space and maintenance due to storage devices like battery banks or flywheels.
3. If the number of distributed generation options is expanded, then the microgrid becomes more complicated to study and analyze.
4. Regulatory issues are more difficult to comply with by the public.
5. Proper care must be taken when a microgrid is resynchronized with the utility grid.
6. The protection system of microgrids is a major challenge for the installation of microgrids [45, 46].

2.12 MICROGRID ARCHITECTURE

Based on the microgrid design concept, the architectures available in microgrids are of three types:

1. AC microgrid
2. DC microgrid
3. AC-DC or hybrid microgrid

The three types of architectures are discussed briefly.

An **AC microgrid** is the main form of a microgrid, and it is similar to a utility grid in terms of operation and stability issues. The basic structure of an AC microgrid (Figure 2.22) is the radial type. Depending upon the nature of the load type, the application of the microgrid and the capacity range, AC microgrids are categorized into three types, system-level, commercial or industrial and rural microgrids.

There is also a need for DC microgrids because the utilization of DC loads has increased in recent years, and there is development in urban distribution grids. As

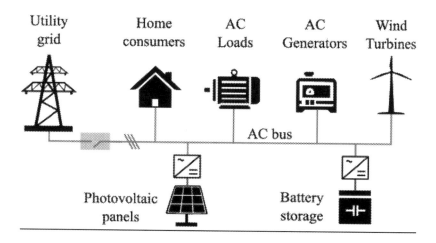

FIGURE 2.22 AC microgrid architecture.

discussed earlier, DC microgrids do not have the problem of synchronization of voltage, as in AC microgrids. This has led to development of DC microgrids at a faster rate. Depending on the type of requirements for electricity, DC microgrids have different architectures, like radial type, ring distribution type and feeder type. The use of power electronic interfaces and bidirectional converters aid the access of AC/DC loads to DC microgrids. Each structure has its own advantages and disadvantages; for example, the ring-type structure provides the advantage of high reliability in power supply; meanwhile, it has a disadvantage of fault detection and hence the protection system is relatively hard to design.

In the **radial type**, different loads of different levels of voltage are connected to a small voltage-rated DC microgrid through a DC transformer. The advantage of the radial type is reliability of fault detection, and hence the design of protection systems is comparatively easy. Meanwhile, the disadvantage is the low reliability in supplying power. The best architecture for rural places and remote regions is the **feeder structure** DC microgrid. Compared to the ring-type structure (Figure 2.23), in a feeder structure, a single bidirectional converter is used to connect the microgrid with the distribution system. The advantages of the feeder structure are the simple construction, and the control system is simpler compared to ring and radial types. Also, fault detection and design of protection systems are simpler compared to ring and radial types (Figure 2.24). Its disadvantage is the poor reliability of the power supply.

Apart from DC load, there are many inductive loads, and in order to supply all the loads with a single microgrid architecture, one has to use an AC-DC microgrid structure (Figure 2.25). The distributed generation and AC or DC loads are accessed by the DC microgrid with the help of a power electronic interface module, and then the AC microgrid gives access to large systems with the help of transformers. The reliability

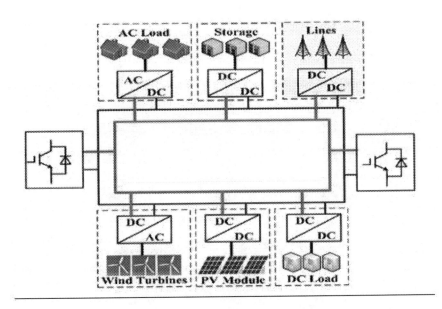

FIGURE 2.23 Ring-type DC microgrid.

Microgrid Design Evolution and Architecture

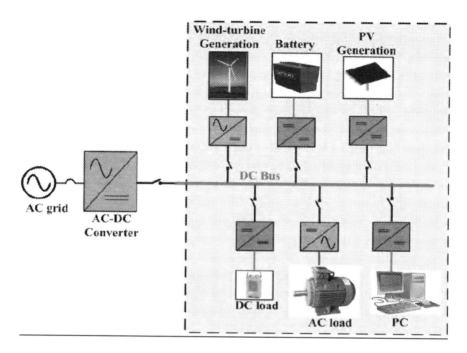

FIGURE 2.24 Radial-type DC microgrid.

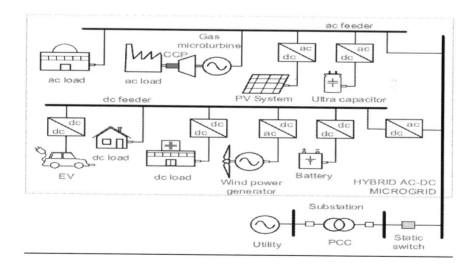

FIGURE 2.25 AC-DC microgrid architecture.

of the power supply, economy, and flexibility and performance requirements can be achieved by an AC-DC microgrid. Since there are large number of inductive loads internally in the architecture of AC-DC microgrids, such as induction motors and transformers, and there are many power electronic devices which introduce harmonics to the system, a suitable static or dynamic reactive power compensation device should be used in the architecture [42–46].

2.13 COMMERCIAL PLANNING OF MICROGRIDS

Microgrids are essential and have numerous benefits in terms of society, but in considering the microgrid as a commercial product, there are many issues to be worked on. From a political perspective, microgrids may not seem useful for local communities because they do not see any sort of benefit in replacing a macrogrid with multiple microgrids. Microgrids are not primary agents for the supply of power, and it might take more time to accept them as primary agents. Utility companies that own the manufacture of transmission wires and transmission components see microgrid entry as competition for their business, so they invest more in improving the reliability of macrogrids rather than supporting microgrids.

Even the integration of microgrids brings a change to existing grid utility codes, which is commercially not feasible. At present, utility companies disagree about accepting and supporting new technology, but unless they release ownership/control of equipment, microgrids will not be available commercially. The entry microgrids brings an industrial revolution, and utility companies might see a drop in their economical status. So, a lot of research is required to make the microgrid a commercial product [47, 48]. A lot of support and resolving methods are required in the context of commercial planning of microgrids.

2.14 CONCLUSION

Research work on the architecture, performance and control methods of microgrid systems is growing more popular with time. Recent developments have proven to establish efficient microgrid systems. In this chapter, the necessity of microgrids and their evolution are discussed so that new ideas and concepts may be developed. The general architecture of microgrids is discussed in brief with every component involved in them. The best alternative for good power quality and uninterruptible power services is the microgrid system. To meet the aim of electrifying each and every house in India, as well as in the world, the development of microgrid systems is a must. It allows access to electricity in remote areas and also encourages the use of non-pollutant energy resources with the help of energy storage technology and other control techniques. The control aspects should be given more importance for a reliable power supply, and also the use of renewable resources as distributed generations creates a chance for sustainable microgrid systems. Stability issues are also discussed in brief, but more research is left to be done on stability issues. Finally, the possible features of future microgrids as a commercial product are discussed, along with political aspects.

REFERENCES

1. E. Hossain, E. Kabalci, R. Bayindir, and R. Perez, "A Comprehensive Study on Microgrid Technology", *International Journal of Renewable Energy Research*, vol. 4, no. 4, December 2014.

2. Yanbo Che, and J. Chen, "Research on Design and Control of Microgrid System", in School of Electrical Engineering and Automation, Tinajin University, Przegląd Elektrotechniczny (Electrical Review), ISSN 0033–2097, R. 88 NR 5b/2012.
3. S. S. Sarita, and S Jayalakshmi, "Hybrid Micro Grid Architectures and Challenges", *International Journal for Modern Trends in Science and Technology*, vol. 2, no. 9, ISSN: 2455–3778, September 2016.
4. H. Karzazi, and R. Lghoul, "Designing a Smart Microgrid", in School of Science and Engineering, EGR 4402–01, 21 December 2017.
5. D. P. Kothari, and I. J. Nagrath, *Power System Engineering*, 3e. New Delhi: McGraw-Hill, 2019.
6. E. Hossain, E. Kabalci, R. Bayindir, and R. Perez, "Microgrid Testbeds Around the World: State of Art", *Energy Conversion and Management*, vol. 86, pp. 132–153, October 2014.
7. "Microgrid Deployment Tracker 4Q12", Pike Research Report, 2012. Sadrul Ula, T. S. Kalkur, Melissa S. Mattmuller, Robert J. Hofinger, Ashoka K. S. Bhat, Badrul H. Chowdhury, Jerry C. Whitaker, and Isidor Buchmann, "The Electronics Handbook", *Second Edition*, April 2005, pp. 1033–1257.
8. Lubna Mariam, Malabika Basu, and Michael F. Conlon, "A Review of Existing Microgrid Architectures", *Journal of Engineering*, pp. 1–8, 2013. doi:10.1155/2013/937614.
9. Changhee Cho, Jin-Hong Jeon, Jong-Yul Kim, Soonman Kwon, Kyongyop Park, and Sungshin Kim, "Active Synchronizing Control of a Microgrid", *IEEE Transactions on Power Electronics*, vol. 26, no. 12, pp. 3707–3719, December 2011.
10. R. Lasseter, B. Schenkman, J. Stevens, D. Klapp, H. Volkommer, E. Linton, H. Hurtado, and J. Roy, "Overview of the CERTS Microgrid Laboratory Test Bed", *PES Joint Symposium Integration of Wide-Scale Renewable Resources into the Power Delivery System*, pp. 1, 1, 29–31 July 2009.
11. N. W. A. Lidula, and A. D. Rajapakse, "Microgrids Research: A Review of Experimental Microgrids and Test Systems", *Renewable and Sustainable Energy Reviews*, vol. 15, no. 1, pp. 186–202, January 2011.
12. Taha Selim Ustun, Cagil Ozansoy, and Aladin Zayegh, "Recent Developments in Microgrids and Example Cases Around the World—A Review", *Renewable and Sustainable Energy Reviews*, vol. 15, no. 8, pp. 4030–4041, October 2011.
13. Ritwik Majumder, "Some Aspects of Stability in Microgrids", *IEEE Transactions on Power Systems*, vol. 28, no. 3, pp. 3243–3252, August 2013.
14. T. Logenthiran, D. Srinivasan, A. M. Khambadkone, and T. Sundar Raj, "Optimal Sizing of an Islanded Microgrid Using Evolution Strategy", in IEEE Conference, 2010, pp. 12–17.
15. S. M. M. Tafreshi, H. A. Zamani, M. Baghdadi, and H. Vahedi, "Optimal Unit Sizing of Distributed Energy Resources in Microgrid Using Genetic Algorithm", in IEEE Conference, 2010, pp. 836–841.
16. F. Tooryan, S. M. Moghaddas-Tafreshi, S. M. Bathaee, and H. Hassanzadehfard, "Assessing the Reliable Size of Distributed Energy Resources in Islanded Microgrid Considering Uncertainty", in WREC, 2011, pp. 2969–2976.
17. Joydeep Mitra, Shahsi B. Patra, M. R. Vallem, and S. R. Ranade, "Optimization of Generation and Distribution Expansion in Microgrid Architectures", in WSEAS Conf. on Power System, 2006.
18. L. Tao, C. Schwaegerl, S. Narayanan, and J. H. Zhang, "From Laboratory Microgrid to Real Markets—Challenges and Opportunities", in 8th International Conference on Power Electronics—ECCE Asia, 2011, pp. 264–271.
19. Jinwei Li, Jianhui Su, Xiangzhen Yang, and Tao Zhao, "Study on Microgrid Operation Control and Black Start", *Power System Conference*, pp. 1652–1655, 2006.

20. A. Salam, A. Mohamed, and M. A. Hannan, "Technical Challenges on Microgrid", *ARPN Journal of Engineering and Applied Sciences*, vol. 3, pp. 64–69, December 2008.
21. W. Jian, L. Xing-yuan, and Q. Xiao-yan, "Power System Research on Distributed Generation Penetration", *Automation of Electric Power Systems*, vol. 29, no. 24, pp. 90–97, 2005.
22. A. M. Azmy, and I. Erlich, "Impact of Distributed Generation on the Stability of Electrical Power System", *Power Engineering Society General Meeting*, vol. 2, pp. 1056–1063, 2005.
23. J. G. Slootweg, and W. L. Kling, "Impacts of Distributed Generation on Power System Transient Stability", *Power Engineering Society Summer Meeting*, vol. 2, pp. 862–867, 2002.
24. R. Lasseter, A. Akhil, C. Marnay, and J. Stephens, et al., "White Paper on Integration of Distributed Energy Resources. The CERTS Microgrid Concept", in Consortium for Electric Reliability Technology Solutions (CERTS), CA, Tech. Rep. LBNL-50 829, 2002.
25. R. H. Lasseter, and P. Piagi, "Control and Design of Microgrid Components, Final Project Report", in PSERC Publication 06–03.
26. Toshihisa Funabashi, and Ryuichi Yokoyama, "Microgrid Field Test Experiences in Japan", Power Engineering Society General Meeting, pp. 1–2, 2006.
27. S. Morozumi, "Micro-Grid Demonstration Projects in Japan", in IEEE Power Conversion Conference, April 2007, pp. 635–642.
28. Oleg Osika, Aris Dimeas, and Mike Barnes et al, "DH1_Description of the Laboratory Micro Grids", in Tech. Rep. Deliverable DH1, 2005.
29. J. Chen, Y. B. Che, and J. J. Zhang, "Optimal Configuration and Analysis of Isolated Renewable Power Systems", in 2011 4th International Conference on Power Electronics Systems and Applications (PESA), 2011, pp. 1284–1292.
30. Eftichios Koutroulis, Dionissia Kolokotsa, Antonis Potirakis, and Kostas Kalaitzakis, "Methodology for Optimal Sizing of Standalone PV—Wind", *Solar Energy*, no. 80, pp. 1072–1088, 2006.
31. S. M. Shaahid, and M. A. Elhadidy, "Economic Analysis of Hybrid PV diesel-Battery Power Systems for Residential Loads in Hot Regions", *Renewable and Sustainable Energy Reviews*, no. 12, pp. 488–503, 2008.
32. J. D. Ren, Y. B. Che, and L. H. Zhao, "Discussion on Monitoring Scheme of Distributed Generation and Micro-Grid System", in 2011 4th International Conference on Power Electronics Systems and Applications (PESA), 2011, pp. 1–6.
33. Y. B. Che, and J. Chen, "Control Research on Microgrid Systems Based on Distributed Generation", *Applied Mechanics and Materials*, vol. 58–60, pp. 417–422, 2011.
34. R. O'Gorman, and M. A. Redfern, "Enhanced Autonomous Control of Distributed Generation to Provide Local Voltage Control", in Power and Energy Society General Meeting—Conversion and Delivery of Electrical Energy in the 21st Century, 2008 IEEE, July 2008, pp. 1–8.
35. J. P. Lopes, N. Hatziargyriou, J. Mutale, P. Djapic, and N. Jenkins, "Integrating Distributed Generation into Electric Power Systems: A Review of Drivers, Challenges and Opportunities", *Electric Power Systems Research*, vol. 77, no. 9, pp. 1189–1203, 2007.
36. R. H. Lasseter, and P. Paigi, "Microgrid: A Conceptual Solution", in Power Electronics Specialists Conference, PESC, June 2004, pp. 4285–4290.
37. N. Hatziargyriou, "Microgrid Control Issues", in *Microgrids: Architectures and Control*, 1st ed. New York: John Wiley & Sons, 2014, ch. 3, pp. 25–80.
38. T. Xu, and P. Taylor, "Voltage Control Techniques for Electrical Distribution Networks Including Distributed Generation", in The International Federation of Automatic Control, 17th World Congress Seoul, 2008.

39. A. Erinmez, D. O. Bickers, G. F. Wood, and W. W. Hung, "NGC Experience with Frequency Control in England and Wales-Provision of Frequency Response by Generators", in Power Engineering Society 1999 Winter Meeting, IEEE, vol. 1, January 1999, pp. 590–596.
40. R. H. Lasseter, "Smart Distribution: Coupled Microgrids", *Proceedings of the IEEE*, vol. 99, no. 6, pp. 1074–1082, June 2011.
41. A. Bidram, and A. Davoudi, "Hierarchical Structure of Microgrids Control System", *IEEE Transactions on Smart Grid*, vol. 3, no. 4, pp. 1963–1976, December 2012.
42. D. E. Olivares, A. Mehrizi-Sani, A. H. Etemadi, C. A. Caizares, R. Iravani, M. Kazerani, A. H. Hajimiragha, O. Gomis-Bellmunt, M. Saeedifard, R. Palma-Behnke, G. A. Jimnez-Estvez, and N. D. Hatziargyriou, "Trends in Microgrid Control", *IEEE Transactions on Smart Grid*, vol. 5, no. 4, pp. 1905–1919, July 2014.
43. M. Chenine, E. Karam, and L. Nordstrom, "Modeling and Simulation of Wide Area Monitoring and Control Systems in IP-Based Networks", in 2009 IEEE Power Energy Society General Meeting, July 2009, pp. 1–8.
44. F. Katiraei, R. Iravani, N. Hatziargyriou, and A. Dimeas, "Microgrids Management", *IEEE Power & Energy Magazine*, vol. 6, pp. 54–65, 2008.
45. A. Vaccaro, M. Popov, D. Villacci, and V. V. Terzija, "An Integrated Framework for Smart Microgrids Modeling, Monitoring, Control, Communication, and Verification", *Proceedings IEEE*, vol. 99, pp. 119–132, 2011.
46. Q. Chen, H. Ghenniwa, and W. Shen, "Web-Services Infrastructure for Information Integration in Power Systems", in Proceedings of the IEEE Power Engineering Society General Meeting, Montreal, QC, Canada, 18–22 June 2006.
47. G. H. Yang, and V. O. Li, "Energy Management System and Pervasive Service-Oriented Networks", in Proceedings of the First IEEE International Conference on Smart Gird Communications, Gaithersburg, MD, USA, 4–6 October 2010.
48. A. Mercurio, A. Di Giorgio, and P. Cioci, "Open-Source Implementation of Monitoring and Controlling Services for EMS/SCADA Systems by Means of Web Services", *IEEE Transactions on Power Delivery*, vol. 24, pp. 1148–1153, 2009.

3 Renewable Energy Source Design in Microgrids

Kumari Namrata, Sriparna Das, R.P. Saini, and Ch Sekhar

CONTENTS

3.1	Introduction	54
3.2	Various Components of Wind Turbines	54
3.3	Different Modes of WECS Operation	56
3.4	Classification of Wind Turbines	58
	3.4.1 Fixed-Speed Wind Turbines	58
	3.4.2 Variable-Speed Turbines	59
	3.4.3 Variable-Speed Turbines with Partial-Scale Converters	59
	3.4.4 Variable-Speed Turbines with Full-Scale Converters	60
3.5	Methodology	61
	3.5.1 Designing a Boost Converter Unit	63
	3.5.2 MPPT Algorithms	66
	3.5.2.1 P&O Technique	67
	3.5.2.2 Incremental Conductance Technique	68
	3.5.2.3 Fixed Voltage Mechanism	69
	3.5.2.4 Current Control Mechanism	70
	3.5.2.5 Temperature Control Mechanism	70
	3.5.3 Design of Inverter Controller Unit	71
3.6	Interfacing Mechanism	74
3.7	Change Regarding Power Generation from Solar PV Cells	76
3.8	Controlling Mechanism	76
3.9	Maintenance Devices	77
3.10	Results and Analysis	78
3.11	Conclusion	78
References		79

DOI: 10.1201/9781003121626-3

3.1 INTRODUCTION

The usage of renewable energy (RE) is becoming the ultimate aim to run industry. Generation of power, along with cleanliness of the environment, is now the prime focus of today's world. The important role of energy in the economic health of any country is shown by the gross national product (GNP). The per capita GNP of any country is correlated to the energy consumption per capita. In the 21st century, the use of renewable energy is one of the most important goals in countries' advancement. Due to easy availability and continuous reform by natural processes, industries are focusing on the usage of renewable sources of energy such as sun, wind, waves, ocean and so on. The initial installation of these sources is a bit difficult during the starting phase, but once it is done, the consumer can get a continuous supply of power with less required maintenance. In this module, a single-grid system is represented by a small microgrid by attaching distributed generation at the distribution or the load side. With the increase in demand of the electrical load, there is an effect of increase in the transmission capacity of long-distance transmission lines, so the dependency of the interconnection of the grid is required. The concept of a microgrid deals with power production in a smaller area. Basically, small generating units are attached to the grid to form a microgrid that will supply a limited region. The extraction of power is done by setting up small generating stations running with energy from sun, wind and other renewable sources of energy. This will reduce the transmission and distribution losses of the system, improving system stability.

In this chapter, the design of renewable energy sources is discussed in detail. Power is extracted from renewable energy sources like wind turbines, solar PV arrays and so on to supply the critical loads connected within the microgrid. Under the presence of a single-grid system, when faults occur, the whole system will have power cut, resulting in a complete shutdown or blackout of the system. In the case of a microgrid connected within a small locality, even if a fault occurs in the system, the microgrid will supply the connected area, preventing shutdown. This is the reason for replacing a single-grid system by dividing it into small interconnected grids. Also, microgrids can store power when a utility grid is present and can supply it back to the utility grid in case of need, for participating in money-earning activities.

3.2 VARIOUS COMPONENTS OF WIND TURBINES

The aim of wind turbine system development is to increase the output power continuously. Today, due to increasing demand for load, the generation requirement is also high. So, the generation capacity is certainly increased for wind turbine systems. This increase in energy is to meet line losses and an increase in load demand. By the year 1999, the average output power from new installations increase to 600KW from wind turbines [1]. To meet the increase in demand every year, production has been increased. In Figure 3.1, the increment in power production in the current scenario is clearly shown.

Generation from wind turbines is classified into two types, onshore and offshore generation. Offshore wind provides a lot of potential. Generation from wind doubled between 2013 and 2016. Wind energy constitutes 16% of the total energy production of electrical energy. The blowing of the wind has kinetic energy and is converted to

Renewable Energy Source Design in Microgrids

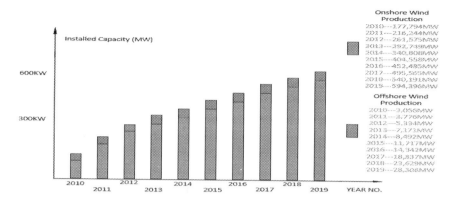

FIGURE 3.1 Installed capacity of power production in corresponding years.

electrical energy by a wind energy conversion system (WECS). The wind rotates the blades of the turbine, and kinetic energy is transformed to rotational energy, rotating the turbine shaft, and coupled with a generator to produce electrical energy. The power generation depends on the size and structure of the blades of the wind turbine. The power output is directly proportional to the diameter of the rotor and proportional to the cube of wind speed.

The major parts of a wind turbine consist of a tower and a nacelle placed on the top of the tower. There are several components of the nacelle, which has a specific function in the energy conversion process from wind to electrical energy. Other main components of a wind turbine include a turbine rotor, gear-box, generator, power electronic devices, control arrangement and transformer, and it is attached to a grid. The kinetic mechanical power from the rotor is transferred to the generator by a transmission system which consists of a rotor shaft, mechanical brakes and gear-box. The mechanical brakes are kept as a backup for sudden stopping of the wind turbine. The main purpose of the gear-box is as a rotational speed increaser. The gear-box converts the slow high torque rotation of the aerodynamic rotor into the faster rotation of the shaft of generator. Considering the geometrical aspects, there are two kinds of gear-box, helical, which consists of gear-wheels with a parallel axis, and the planetary type consisting of epicyclic trains of gear-wheels.

For variation of speed and torque, a gear-box is present, making the system bulky. As a result, in modern wind turbine technologies, the generator is designed with a multi-polar structure in order to adapt the rotor speed to the generator speed. The speed of the generator decreases by increasing the pole pairs, and thereby the system becomes lighter. The rotor of the wind turbine captures the power from the wind and converts it into mechanical power [2]. The aerodynamic rotor is made up of a hub and blades, and the blades are attached to the rotor using mechanical joints.

The generator acts as an electromechanical component in converting mechanical to electrical power. The generator typically consists of a stator and a rotor. The stator is static, consisting of coils mounted in a certain fashion. The rotor is the rotating

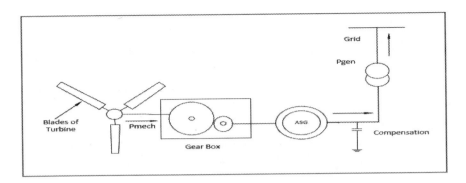

FIGURE 3.2 Block diagram model of a typical wind turbine.

part of the generator and generates the magnetic field. By the rotation of the rotor, the magnetic field passes through the stator windings and voltage is induced in the terminals of the stator. Either a permanent magnet or electromagnet can be used in the rotor. When the magnetic field of the stator follows the magnetic field of the rotor, the generator is termed a synchronous generator; otherwise, it is called an asynchronous generator. A synchronous generator can be used or replaced by an induction generator, which is an asynchronous-type generator. In Figure 3.2, a model of a wind turbine is seen. The diagram depicts from the rotation of the blades of the turbine to the end connection with the grid. The presence of the transformer is to step up the voltage, as the voltage generated by the turbine is lower.

3.3 DIFFERENT MODES OF WECS OPERATION

A synchronous generator (SG) operates at synchronous speed. It is dictated by the frequency of the connected grid, regardless of the torque applied [3]. The magnetic field in the synchronous generator can be created by using permanent magnets or conventional field windings. The speed at which the generator rotates is dependent on the frequency of the rotating field and the number of pole pairs of the rotor. It is much more expensive than an asynchronous generator. However, one significant advantage in comparison to an asynchronous generator is that it does not need a reactive magnetizing current, and thus no further compensation is required. There are two types of synchronous generators:

1. Wound round synchronous generator
2. Permanent magnet synchronous generator

Wound round synchronous generator (WRSG): The WRSG is considered the workhorse of the electrical power industry. Its stator windings are attached to the power grid; hence, the speed of rotation is fixed, as the frequency of the supply grid remains fixed. DC current flows through the rotor winding to generate the exciter field, which rotates with synchronous speed.

Permanent magnet synchronous generator: Here the stator used is of the wound type, and a permanent magnet is present along with the rotor. As there is an absence of an energy supply, its efficiency is high. But the cost of producing this type of magnet is high, and it is also a bit difficult. If permanent magnets are used, then converters should be present to take care of voltage and frequency.

Asynchronous generator: There are various advantages to using an asynchronous generator. This includes the simplicity of their working and formation, robustness and low cost. But the main disadvantage is due to the requirement of reactive magnetizing current by the stator. As permanent magnets are not present and separate exciting is not done, for excitation, reactive power consumption is necessary. The reactive power will either be provided from the grid or by introducing new power electronic devices in the system. The electric field is induced by making a relative motion known as a slip between the rotor and stator field, which is rotating in nature. This results in the formation of torque that acts on the rotor. The rotor includes two basic types of design, the squirrel cage and the wound rotor.

In the case of a squirrel cage induction generator, there are slots placed on the rotor, and end rings are attached on both ends. It is robust in nature, and the speed varies much less with changes in slip with the speed of the wind. For wind turbines using this type of generator, the presence of soft starting and reactive power compensation equipment is mandatory. The slope of the curve produced by the torque to speed variation of the generator is very high, so any type of change in wind speed directly affects the grid [4]. This creates a major problem when the wind turbine is synchronized with the grid, as incoming current increases up to seven to eight times the rated value. In the case of a weak grid, these currents will cause a very high amount of voltage fluctuation. That is the reason a soft starter is mandatory to limit the current and to connect with the grid. The reactive power that is generated is also dependent upon the wind. When the wind is blowing at high speeds, the wind turbine is able to produce much more active power if and only if the supply of reactive power to the generator is high. If it is low, then the reactive power is drawn from the power grid, and transmission loss will occur and the grid might become unstable. To prevent this situation, converters and capacitor banks are used for reactive power compensation.

In the case of the wound round induction generator (WRIG), rotor windings are connected with power electronic equipment or by an arrangement of slip rings and brushes. In this way, the electrical parameters can be controlled from outside. With the help of power electronic devices, power can be taken from the rotor circuit, and magnetizing of the generator can easily be done from the rotor to stator circuit. The main disadvantage is that it is not at all robust and is very expensive. The two types of wound rotor induction generators used in industries are opti slip IG and doubly fed IG.

Opti slip IG: These were widely used in the 1990s, where resistance is added to the rotor winding and this resistance is of variable type. This can be controlled optically by the converter placed on the rotor winding. By this arrangement, brushes and slip rings can easily be removed. But the controlling speed is dependent on rotor resistance [5].

Doubly fed IG: The stator winding is connected to a power grid, so the frequency will not change and will remain the same as the grid frequency. The rotor is attached to the grid via power converters. The speed range varies according the size of the converters used. Full speed is not required, so converters are designed for 70% of the speed to be utilized. This selection of converters depends upon cost. So, when the speed range increases much more from synchronous speed, the cost increases [6]. It can handle active as well as reactive power components.

3.4 CLASSIFICATION OF WIND TURBINES

Wind turbines are classified by how they control the speed. They are as follows:

1. Fixed-speed turbine
2. Variable-speed turbine
3. Variable-speed turbine with converters (partial scale)
4. Variable-speed turbine with modified converters

3.4.1 FIXED-SPEED WIND TURBINES

A fixed-speed turbine (Figure 3.3) uses a gear-box and squirrel cage IG to convert the mechanical form to the electrical form of energy. This concept was also found in many Danish wind turbines [2] operated in the 1980s; therefore, it is referred to as "Danish concept." Here the transformer is placed before the grid to step up the voltage to the required level. Soft starters are present for synchronizing between the grid and wind turbine, whereas the presence of the capacitor is for reactive power compensation.

Since the squirrel cage induction generator operates in a small region with a smaller deviation from synchronous speed, this type of wind turbine runs at a constant speed. It is simple and robust in nature. The cost of the design is relatively low. As it operates in almost constant speed, in the presence of fluctuations in wind speed, there will be a change in mechanical power, effecting a change in electrical power and reducing the system stress [7]. But the major disadvantage is the limitation of the control mechanism to look into the power quality of the system, and reactive power consumption cannot be controlled in this type of wind turbine system.

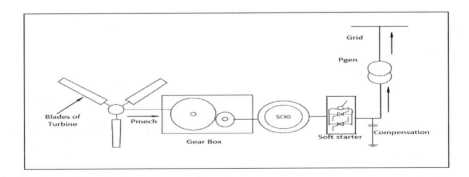

FIGURE 3.3 Structure of fixed-speed wind turbine.

3.4.2 VARIABLE-SPEED TURBINES

This configuration involves limited control of the speed of the wind turbine, known as opti slip or flexi slip [5]. There is a wound round induction generator where the stator is directly connected to the power grid. There is variable resistance, connected in series with the rotor winding. By varying the resistance, the speed and energy generated can easily be controlled. Due to the presence of resistance in the configuration, losses occur due to heat. A soft starter is present to control the in-rush current, and a reactive compensator supplies reactive power to the generator. The control arrangements include controlling of variable resistance and pitch [8]. The main advantage is that, without the presence of slip rings, the operating area of the machine can be increased by varying the speed. But there are certain limitations also, such as a speed limitation due to dependency on rotor resistance. Power loss also occurs due to resistance in the form of heat and poor control of active and reactive power. Figure 3.4 depicts the model of this type of turbine.

3.4.3 VARIABLE-SPEED TURBINES WITH PARTIAL-SCALE CONVERTERS

This type of configuration resembles a variable-speed wind turbine using DFIGURE-based generation. The stator of the WRIG is connected to the power grid directly, and the rotor of the WRIG is connected to partial-scale back-to-back converters via slip rings. The speed varies over a wide range. A partial scale signifies 30% of the power generated. The rating of the converters determines 30% deviation from the synchronous speed. The turbine decouples the mechanical and electrical system frequency and performs variable-speed operation. The reactive power is compensated, and there is synchronization between the grid and the wind turbine. There is no requirement for a soft starter. The slip is placed to transfer rotor power with the help of a partial scale converter. The converter controls the rotor speed by controlling its frequency and allowing variable-speed operation. It also controls active and reactive power.

The control of the turbine blade by pitch control and converter by electrical control resemble the control arrangements. It is a fairly expensive method, but in previous methods, the rotor energy is wasted through the use of resistance. However, in this mechanism, this energy can be fed to the grid with power electronic devices. It can

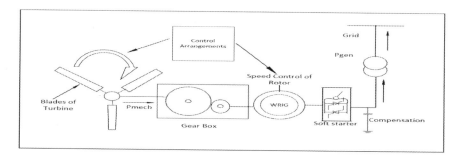

FIGURE 3.4 Model of variable-speed wind turbine.

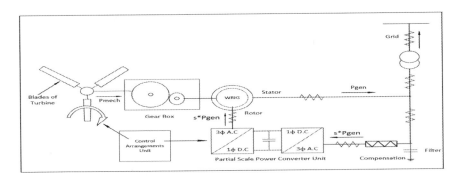

FIGURE 3.5 Model of partial scale converter-based wind turbine.

operate over a much larger area due to huge variations of speed, while the size of the converter is dependent on the variation in speed, and losses in the converter are also much less. The main disadvantage is that converter and slip rings are present, which require maintenance, and occurrence of faults in the grid due to non-maintenance may lead to failure of the machine. In Figure 3.5, an overview of a converter-based wind turbine is shown.

3.4.4 Variable-Speed Turbines with Full-Scale Converters

In this case, full-scale converters are used where the speed range can vary up to 100%. All other advantages are discussed in the case of a partial scale converter using a wind turbine. This can be implemented by using an electrically excited DC generator (WRSG or WRIG) or permanent magnet synchronous generator (PMSG). It doesn't have a gear box; multiple pole generators are used instead. This will have low speed, as the generator is connected to the hub of the rotor directly. But higher torque is required, so the number of generator poles can be increased to meet the demand. This has a huge advantage because of high efficiency, no slip rings and no gear-box so the system becomes less bulky, has better speed range support by the grid, is low cost and reliability also increases. In Figure 3.6, a full-scale converter-based wind turbine model is shown.

In Figure 3.7, there is clear evidence that by varying the wind speed, the power production from the turbine is varied. When the wind speed is low, the extraction of power from the wind by the blades of the turbine is low; hence, there is less production of energy. With increasing wind speed, energy production increases up to a certain limit. The limit is known as the Betz limit, up to which the energy-producing process can be performed. If we increase past the Betz limit, then the wind speed forces the turbine blade to stop rotating, and shutdown of the generation process will occur. This is clearly shown in the figure. The red line is the power generation line, that is, how much power is produced with changing wind speed, and after rising to a certain limit, it becomes constant. That point is the maximum capacity of the wind turbine.

Renewable Energy Source Design in Microgrids

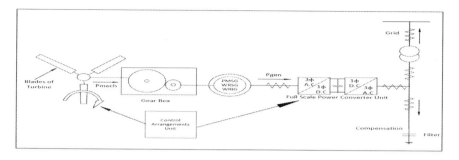

FIGURE 3.6 Full-scale converter-based wind turbine model.

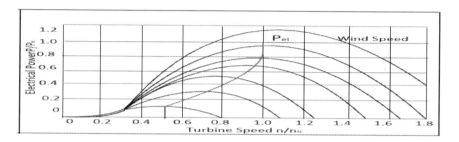

FIGURE 3.7 Effect of changing wind speed on output of the industry [9].

3.5 METHODOLOGY

A simple wind turbine model is taken whose inputs are generator speed, pitch angle and wind speed. The wind turbine rotates for production of energy so the output is torque and mechanical power, as shown in Figure 3.8. This torque helps to run the permanent magnet synchronous generator. The use of the permanent magnet synchronous generator is implemented for certain conditions as follows:

1. Using DFIGURE-based generation is becoming old fashioned nowadays. To control high speed, a gear box is needed, which makes the system more costly. PSMG overcomes this, as it is related to the low-speed process, so a gear box is not required, making the system light, minimizing the cost and also improving the efficiency of the system.
2. Due to the presence of permanent magnets, there is no need for electrical energy supplied from external sources. Therefore, no copper loss will take place within the exciter. There are also no mechanical arrangements such as brushes and commutators, so mechanical losses will not take place, making the system more compact than earlier.
3. In rotor winding, there is no high-powered magnet, making the size small.
4. As the rotor circuit is not present, no current will flow through it; therefore, losses are minimized, as no heat is produced in the rotor circuit. Heat is

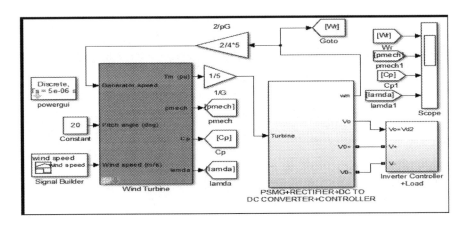

FIGURE 3.8 Simulink model of WECS system [9].

produced only in stator winding heat, and, as it is static and the outermost part of the machine, heat can easily be released through the air, and noise production is also less.

For these advantages, the use of PMSG-based generation is essential compared to other models [10]. When the generator is made to run by supplying torque, it helps in the formation of three-phase voltage at the output terminal. Here voltage and current are measured by using a measurement block in MATLAB, and these three phases are converted to a single-phase supply by using a rectifier unit or universal bridge in MATLAB.

To step up the voltage to the required, a level single-phase output of the rectifier unit acts as an input to the boost converter. By using this method, the generation is done. Only certain parameters are known, including supply frequency, which will remain constant throughout the process, and the input voltage magnitude.

By considering the output current per requirements, the design of resistance, inductance and capacitance, which are the parameters of the converter units, are designed. There might be ripples at the output voltage and current waveforms, so the capacitor is added before taking the output to remove the ripples and lessen the total harmonic distortion, which improves the power quality. So, the function of the capacitor is to remove all the distortions and ripples that are present in the output voltage and current waveforms. Here the nature is single phase, but as the required loads are three-phase in nature for being beneficial, conversion from single to three-phase is done by introducing an inverter block and designing an inverter controller unit. After the running of the PMSG, the mechanical speed, which acts as an output of this block, is fed back as an input to the turbine block by adding a gain in its path to continuously run the process. This model acts as an example of closed-loop feedback system, as the output of the generator enters as input to the turbine.

The generator, after getting into motion, controls the speed of the wind turbine. The entire simulation is done step-by-step, and after performing each block separately

Renewable Energy Source Design in Microgrids

and simulation is done, they are attached to get the final result of the wind turbine. In the following simulation, there are three blocks: the wind turbine block, which is an inbuilt block and no changes are made in it; the block that contains the part starting from input, taking to the PMSG to the output of the boost converter; and last is the inverter block, which takes the input from second block and generates three-phase voltage. These three blocks denote three sub-systems, and finally they are linked together to run the whole simulation [11]. In Figure 3.8, there is a clear view of the entire model from outside.

3.5.1 Designing a Boost Converter Unit

To increase the voltage to the desired level, a boost converter is attached into the system. The generation capacity is always less compared to the demand. So, the design of the boost converter comes into play. This follows a DC-to-DC conversion scheme. Due to the increase in load and limited generation capacity, this converter is used to meet the demand of the system. The design of the converter deals with setting pulses or the firing angle of the switch. The switch should have the minimum amount of losses. Another alternative to replacing the converter is using cells in series-parallel fashion until the desired output is reached.

To design this converter, two parameters are essential, the output or the desired voltage and the load current [12]. The design includes setting the values of the elements such as the inductor, capacitor and resistor that are used in this boost topology. The current and voltage can be considered per requirements. The frequency of the system while designing the whole model is fixed. If the frequency of the module is varied, there will be disturbance introduced into the system and the system will lose synchronism. Considering all the mentioned parameters, the design of this converter section is shown in Figure 3.9.

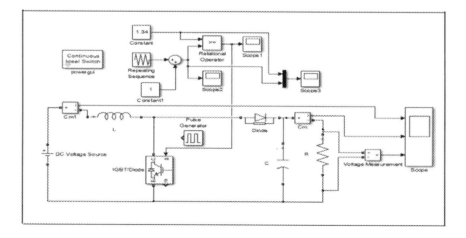

FIGURE 3.9 Boost converter circuital view.

The switch with minimum losses that is preferred here is insulated gate bipolar transistor(IGBT). Pulses are generated by comparing a constant value with repeating sequences. The relational operator compares those values and sets the pulses for the switch. The width of pulse considered here is 67%, and no delay is introduced in the system. There is another alternative while designing the pulses. The introduction of a pulse generator will serve the purpose directly. In this case, a non-positive part will come into the scenerio; to eliminate this, adding a costant term is done, as a pulse can never be in a negative sequence. The various formulas [13] that are used for designing the converter are as follows:

$$V_0 = \frac{V_{in}}{(1-D)} \tag{3.1}$$

$$\text{Load Resistance} = \frac{V_0}{I_0} \tag{3.2}$$

$$\text{Duty Cycle} = 1 - \frac{V_{in}}{V_0} \tag{3.3}$$

$$\text{Capacitance}(C) = \frac{I_0 D}{f_s V_s} \tag{3.4}$$

$$\text{Inductor}(L) = \frac{V_s D}{f_s I_0} \tag{3.5}$$

Thus, the design of the generated voltage is done in the stated fashion. In Figure 3.10, it is clearly shown how the PMSG [9] is attached with the rectifier unit followed by the boost converter.

A new block known as the MATLAB function block can be seen in Figure 3.10. The intention of getting the output constant is prescribed in this block. To get the maximum output, the use of maximum power point tracking (MPPT) is necessary. The code is written in that function box. It generates the pulses for the switch that is the IGBT/diode used in the converter section. Here, a perturb and observe algorithm is written, which is explained subsequently. The output of the increased voltage and current from the boost converter is fed as input to the inverter block, and the waveforms of these parameters of the converter are shown in the following.

In Figure 3.11, it is seen that the output voltage increases up to a certain limit, and then after 0.2 seconds, the voltage becomes constant and the value reaches 34V. The boosting is done from 0 to 34 V, which will be fed as input to the inverter. As the voltage remains constant, so it easily shows the MPPT circuit is working accurately; otherwise, there will be fluctuation in the voltage waveform. Next, the current waveform is also necessary, and how it varies is shown in the following.

In Figure 3.12, it is observed the current also varies in the same way as the voltage and reaches up to a certain value, then falls a bit and becomes constant. The current increases up to 0.025 seconds, and when it reaches 0.2 seconds, it becomes constant, MPPT is verified [14] and the load attached is resistive, as the current is in phase with

Renewable Energy Source Design in Microgrids

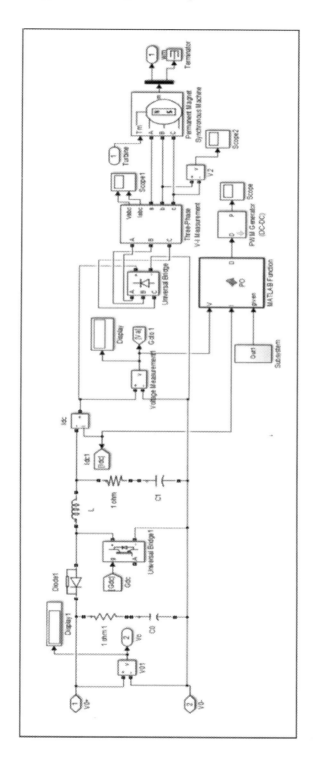

FIGURE 3.10 Connection of the rectifier generator, rectifier and converter units.

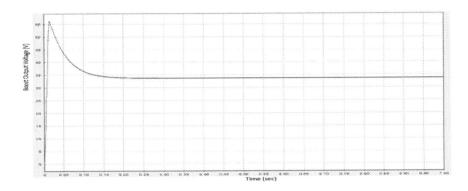

FIGURE 3.11 Voltage variation of the converter section.

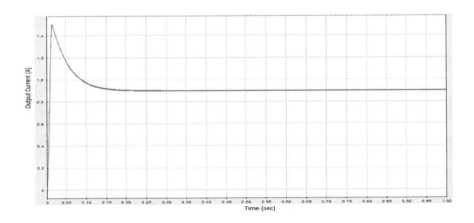

FIGURE 3.12 Current variation of the converter section.

the voltage. If the load changes, there will be a phase difference that can be observed between those waveforms. The maximum value of the current is 1.5A; it can be made higher if we design the boost converter for a greater range of values.

3.5.2 MPPT Algorithms

The use of this algorithm is to run the system at the highest efficiency so that the maximum amount of power can be extracted from the system. The dealing power of a DC-to-DC converter is quite low, so any loss occurring in the system will cause a decrease in system efficiency. So, the system is to run at MPP such that pulses are generated or maintained in such a way that the system follows the MPP track. There are various mechanisms followed while performing this tracking process. The design of this algorithm is associated with the converter on which the tracking is to be done to obtain the maximum amount of power. The various methods of MPPT are categorized as follows:

1. Perturb and observe (P&O) technique
2. Incremental conductance technique
3. Fixed voltage mechanism
4. Current control mechanism
5. Temperature control mechanism

3.5.2.1 P&O Technique

The use of this algorithm is the most common. In this mechanism, searching is not dependent on external parameters like atmospheric condition, humidity and so on. The parameters that are necessary for this search are only current and voltage. The tracking of MPP is done by comparing the final and previous sets of values. By comparing these values, pulses are generated for the switches of the converter. When a perturbation is inserted into the system, the on and off periods of the system are affected, so if the nth value is higher than its previous state value, the duty is increased, as the MPP is towards the right side, and if the checking fails, then the duty is decreased. By this mechanism, the tracking process is performed. A diagrammatic representation of this algorithm [9, 15] is given in Figure 3.13.

The MATLAB function used in Figure 3.10 is shown in Figure 3.14 [16].

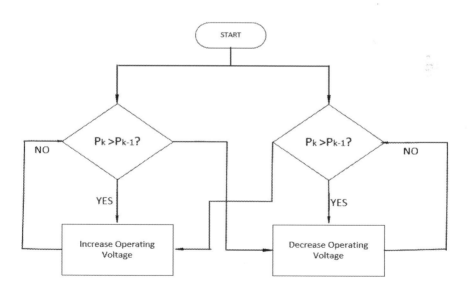

FIGURE 3.13 Perturb and observe algorithm.

```
function D = PO(V, I, given)
Dini = given(1);                Function Initialisation
Dll = given(2);
Dul = given(3);
dD = given(4);
persistent V1 P1 D1;

if isempty(V1)
    V1=0;
    P1=0;
    D1=Dini;
end

P = V*I;
dV = V - V1;                    Searching for maximum
dP = P - P1;                         power point
if dP ~= 0
    if dP < 0
        if dV < 0
            D = D1 - dD;
        else
            D = D1 + dD;
        end
    else
        if dV < 0
            D = D1 + dD;
        else
            D = D1 - dD;
        end
    end
else
    D = D1;
end
if D >= Dul || D <= Dll
    D = D1;
end
D1 = D;          Storing the value after being
V1 = V;                   searched
P1 = P;
```

FIGURE 3.14 Algorithm used in MATLAB function.

3.5.2.2 Incremental Conductance Technique

It is most useful when it is attached with perfection and speeds. Here, only current and voltage parameter are needed [17]. So, from this, power is generated and the derivative of power with respect to voltage, that is, $\frac{dP0}{dV0}$, is done. $\frac{dP0}{dV0}$ can be written as:

$$\frac{dP0}{dV0} = I0 + \frac{dI0}{dV0} = I_{n0} + \frac{I_{n0-1} - I_{n0}}{V_{n0-1} - V_{n0}} \tag{3.6}$$

Renewable Energy Source Design in Microgrids

There are three conditions while finding the MPPT:

1. $\frac{dP0}{dV0}$ is greater than zero, which means the point is on the left side of the maximum power place, so the duty must be increased to get the maximum power point location.

2. $\frac{dP0}{dV0}$ is less than zero, which means the point is on the right side of the maximum power place, so the duty must be decreased to get the maximum power point location.

3. $\frac{dP0}{dV0}$ is equal to zero, which is the point where maximum power can be extracted, so the duty is kept fixed.

By following these three conditions, the point can be identified.

This process is costlier than the first one, as it requires two sensors at the same time. It is also far more complex in design aspects, and there are fewer variations taking place at the point of maximum power. Its parameters are also not dependent on external factors such as surrounding, climate and so on.

3.5.2.3 Fixed Voltage Mechanism

In this method, the power gets varied with change in voltage due to effect of change in temperature and solar irradiance, keeping one fixed at a time is the prime idea. With increasing in the amount of incoming sun to the PV cell, more power will be produced if the temperature is kept constant, and the voltage also increases slowly, as shown in Figure 3.15. On other hand, if the radiation from the sun is assumed to be constant and the temperature is increased, then the production capacity of power will increase for the PV cell, but it will be seen that with the increase in temperature, there is reduction in voltage. With this mechanism, it is easy to track and works if only a single parameter is available. In Figure 3.16, the effect of the change of power with voltage can be noted with the change in solar rays and temperature [18].

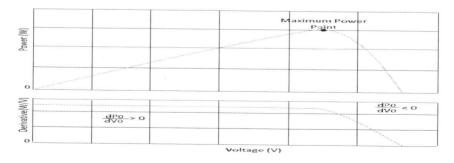

FIGURE 3.15 Output of PV cell when incremental conductance is applied.

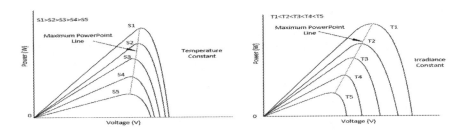

FIGURE 3.16 Effect of change in power with the change in temperature and solar insolation, varying one at a time.

3.5.2.4 Current Control Mechanism

In this, an I-V (current versus voltage) graph of the solar cell is required. The current value at which MPP occurs is related to the short circuit current by a constant value K_{sc}. This constant can operate in a wide range of solar insolation and temperature. Generally, the value of this constant for PV cells is 0.92. But the disadvantage is that the short circuit current must be measured within some interval of time, but its measurement is a bit difficult, as during short circuit, there is no flow of power within the loads.

3.5.2.5 Temperature Control Mechanism

In this mechanism (Figure 3.17), the temperature analyzer is replaced by a current analyzer. This model is formulated from the relation of voltage dependent on the amount of sunlight that falls on the PV cell or its temperature [19]. The formula is shown as the following:

$$V_{mppt}(t) = V_{mppt} \times t_{reference} + U_{mppt}(t - t_{reference}) \qquad (3.7)$$

This shows that the maximum power point voltage at any time t will be equal to the maximum power point at a reference time in addition to the temperature coefficient. There is also a separate formula for the calculation of duty that is given by:

$$\Delta(Duty) = [V_{output} - V_{mpp}(t)] \times K \qquad (3.8)$$

Using formula of the duty cycle, it is found, and the comparison is done as shown in the flowchart. Here, K resembles the size or the value of how much increment or decrement is done, or the step size, and the use of $\Delta(Duty)$ is to find the point of maximum power. This process is very slow, so it is not used as frequently as the P&O mechanism.

By these methods, the tracking of that point is done. Because there are fewer parameters needed, tracking can be done efficiently depending upon step size and independently of any external parameters like climate and season. This adds more benefit for users to use the P&O algorithm (Figure 3.18), so the authors have used this method to track the MPPT in the case of power generation from wind turbines.

Renewable Energy Source Design in Microgrids

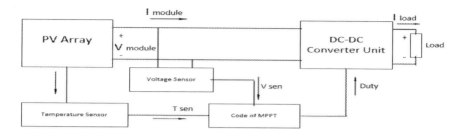

FIGURE 3.17 Overview of temperature control mechanism.

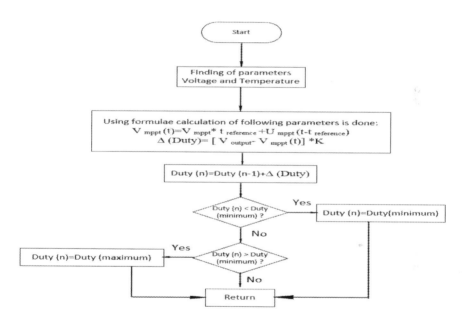

FIGURE 3.18 Algorithm of temperature control method.

3.5.3 Design of Inverter Controller Unit

The loads attached to the distributor end are not a single-phase supply; they require a three-phase supply, so the design and control of the inverter block are mandatory. In design, the theme is current control, so the voltage source PWM technique is followed. The use of the controller unit is to make a small deviation in the load current and to keep the direct voltage constant. In Figure 3.19, it is seen that two loops have been introduced while designing this model. One will take care of current and the other voltage. The output voltage is compared with the reference voltage, and an error is generated, and via a multiplier it is changed and again fed back to the system [20]. The output of the converter acts as input to the inverter block and is attached to the load, which follows the requirements from the customer.

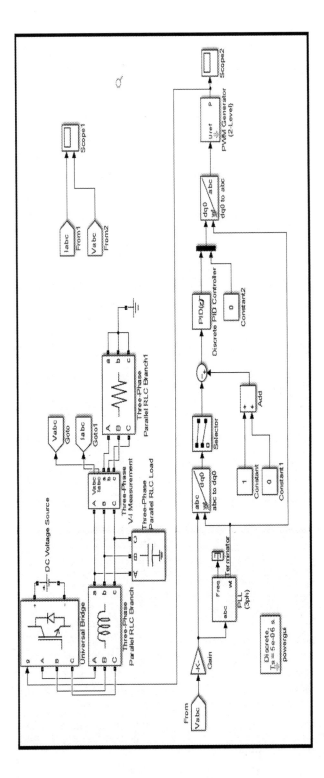

FIGURE 3.19 Design of inverter controller unit [9].

Renewable Energy Source Design in Microgrids

FIGURE 3.20 Three-phase output waveform of inverter unit.

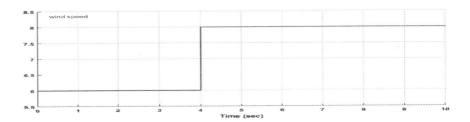

FIGURE 3.21 Wind speed variation with time.

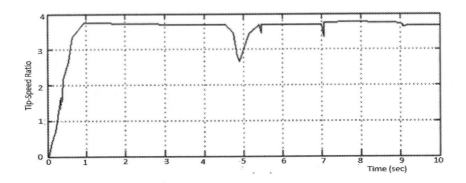

FIGURE 3.22 Variation of tip-speed with change in time.

The design of the inverter module and the three-phase output waveforms of the current and voltage are shown in Figures 3.19 and 3.20, respectively.

The wind speed, which is given as an input to the turbine block, is shown in Figure 3.21, the final variation of tip speed with the change of time is shown in Figure 3.22 [9], and the PMSG per phase output waveform is shown in Figure 3.23 [21].

3.6 INTERFACING MECHANISM

A generator is used to convert mechanical to electrical energy and is connected to a power grid through power electronic devices. Being placed between the wind turbine and power grid, it should satisfy the requirements of both ends for cost minimization and easier maintenance. On the generator end, maximum power extraction is the main intention due to flowing of the wind and rotation of the turbine blades. On the grid end, the power electronic devices must have synchronization with the grid even when there is a change in wind speed; these converters help maintain the system frequency and output voltage and also maintain reactive power.

Renewable Energy Source Design in Microgrids

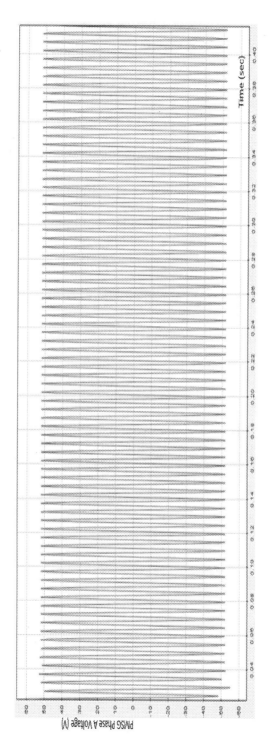

FIGURE 3.23 Single-phase output voltage of PMSG.

3.7 CHANGE REGARDING POWER GENERATION FROM SOLAR PV CELLS

To generate power from the sun's rays, there is a change in the mechanism of the wind. In the case of generation of power from wind, wind turbines are used along with PMSG. The blades of the turbine, on contact of the wind, absorb the energy from the wind and convert it into rotational energy to rotate the shaft of the PMSG, and this generator helps in the generation of the three-phase balanced sinusoidal waveform. But, in the case of solar power generation, there is a replacement of these two blocks with only multiple PV cells to form a PV array. In the case of a standalone system [22], where there is no connection to a utility grid, PV cells are connected to a charge regulator unit, followed by a battery and inverter and finally loads are connected, while in the case of a hybrid system [23], diesel generators and the PV cells produce heat, and that energy is used to run the generator. By proper controlling and handling of power, it is supplied to the loads, forming a hybrid system. The last type is a grid-connected system in which the PV cells absorb energy from the sun, and with proper controlling and handling of power, they are connected directly to the grid. So, in the case of a grid-connected system, the converter and inverter designs are similar to wind turbine power generation, but the main difference is that the turbine and PMSG are replaced by a PV cell. The various types of power production from PV cells are shown in Figures 3.24 and 3.25, where the first one shows a standalone system and the second shows hybrid and grid-connected systems.

3.8 CONTROLLING MECHANISM

The extraction of maximum power is the ultimate goal from the wind turbine, so it is associated with control arrangements for making wind turbines operate within a stable region. There are two types of controlling methods:

1. Passive control
2. Active control

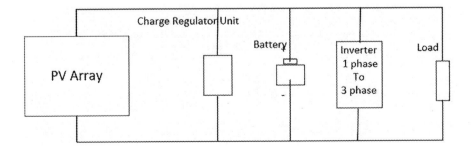

FIGURE 3.24 Standalone PV system.

Renewable Energy Source Design in Microgrids

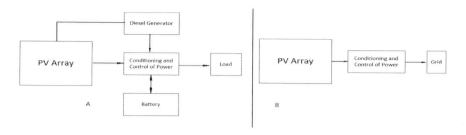

FIGURE 3.25 (A) Hybrid PV system, (B) grid-attached system.

Passive control uses its own means and takes help from the forces of nature to control the parameters, whereas active control uses electrical, mechanical or other means to control the parameters. When the wind speed is low, the control design should operate in generating optimal power and making the wind turbine run at maximum efficiency, while when the wind speed is high, the wind turbine should be within the limits of producing rated power by reducing the force acting on the blades and loads [24].

Passive or Stall Process: It is a very simple method of power control. The blades of the turbine are fixed to the hub such that the wind hits at a specific angle. So, the rotor is made in such a way that it stalls itself to lower the efficiency when the wind speed reaches the upper limit, that is, the rated limit.

Active Process: The other name for this method is the pitch control method. The rotation of blades, either fast or slow, corresponds to high or low generation of power. In the case of active control, the geometry of the rotor needs to be changed, so it's a bit expensive. The system is designed by using an electronic or computer system controlling the motor. It is controlled by varying the pitch angle, making a small deviation with respect to time when the speed of the wind changes [25]. By this process, only output power is maintained.

3.9 MAINTENANCE DEVICES

The devices that are used in maintaining synchronization are as follows:

1. Soft starter
2. Capacitor bank
3. Frequency converters

Soft Starter: helps in limiting in rush currents so disturbance to the system is minimized. Without it, the current becomes seven to eight times the rated value, causing a break in synchronization with the grid and making the system unstable [26].

Capacitor Bank: helps in providing reactive power to asynchronous-type wind turbines that help lessen the reactive power taken from the grid by the asynchronous generator [27].

Frequency Converters: help in maintaining the frequency and voltage of the generator and connecting various electrical systems with different frequencies. It is also called an adjustable speed drive, in which the rectifier units are present for converting AC supply to direct form. Capacitors used for storage purposes and inverter unit need AC supply because most of the loads are AC in nature, so the inverter converts DC to AC, and the AC current flows through the load and is distributed to the consumers according to their needs [28].

3.10 RESULTS AND ANALYSIS

The entire process of how power can be extracted from a wind turbine and phase transformation is shown in this chapter. Each model was separately simulated in MATLAB 2015 a, and the waveforms are shown subsequently. The wind speed varies from 6 to 8m/s, as shown in Figure 3.21, and how the power is extracted depends on the variation with wind speed, shown in Figure 3.7. To obtain the maximum amount of energy, the authors used the P&O mechanism of MPPT and explain the entire process with the flowchart in Figures 3.13 and 3.14. There are certain values considered in running the simulation. The density of air is assumed to be constant, ρ=1.08m³/kg; the value of the gear ratio is 5; the radius is 1.525m; and the speed is 20 revolutions per second, which are the basic parameters in terms of the wind turbine block and control part. Next, comes the design of the permanent magnet synchronous generator in which stator phase resistance is considered to be 1.5 ohms, armature inductance 10mH, flux linkage 0.119Wb, inertia 2J (kg.m²) and the number of poles four. A three-phase round rotor is used. In the case of the design part of the boost converter section, two resistors of 1ohm, C_0=3500 microfarad, C_1=500 microfarad and inductance of 200 microhenry are used. The switching frequency is considered to be 5000Hz throughout the system. The block diagram model is shown in Figure 3.8, and the design of the boost converter is depicted in Figure 3.10, followed by its voltage in Figure 3.11 and current in Figure 3.12. The design of the inverter controller to change to three-phase was discussed, and its design is shown in Figure 3.19 and three-phase waveform of current and voltage in Figure 3.20. From the experiment, it is concluded that with an increase in the speed of the wind turbine, the power extraction also increases, but up to a certain limit, and wind turbines should operate only within than limit, known as the Betz limit. The variation of the tip speed ratio can be seen increasing up to 2 seconds and after that attains a constant value. All the values considered while performing this model were disclosed in the case of the wind turbine model, and, further, this chapter discussed how and what replacements need to be done in the case of solar implementation if it is replaced by a wind energy system.

3.11 CONCLUSION

In this chapter, it is clearly seen how the design of renewable energy sources [29–30] is taking place. If a fault occurs in a single-grid network, then there will be a complete shutdown, but this can be prevented by breaking a single grid system into microgrids. So, the design of renewable energy sources is very important. This chapter clearly shows each and every model simulated in software, its waveforms and the process of how this implementation [31–32] is done along with its waveforms in a practical way. There are many renewable energy sources, but this chapter focuses on solar and wind power generation, which are considered the basic sources of supply. To obtain maximum power, MPPT methods [33] are also briefly explained. A detailed theory of every model along with its waveforms is clearly explained and discussed in this chapter.

REFERENCES

1. "Report on Renewable Energy," International Renewable Energy Agency (IRENA). [Online]. www.irena.org/wind.
2. S. Muller, M. Deicke, R. W. De Doncker, "Doubly Fed Induction Generator System for Wind Turbines," *IEEE Industry Applications Magazine*, vol.8, no-3, pp. 26–33, May 2002.
3. Anca D. Hansen, "Chapter 8: Wind Turbine Technology" in *Wind Energy Engineering, Part. 3 Wind Engineering*, vol.26, pp. 145–160, 2008.
4. Varin Vongmanee, "Emulator of Wind Turbine Generator Using Dual Inverter-Controlled Squirrel Cage Induction Motor," in 2009 International Conference on Power Electronics and Drive Systems (PEDS), 2–5th November 2009, pp. 1313–1316.
5. K. Vinoth Kumar, A. Nageswar Rao, M. P. Selvan, "Mitigation of Output Power Fluctuations in Wind Farms with Opti-Slip Induction Generator," in 2009 IEEE PES/IAS Conference on Sustainable Alternative Energy, 28–30th September 2009, pp. 1–5.
6. Mohamed Hallak, Mourad Hasni, Mohamed Meenaa, "Modelling and Control of a Doubly Fed Induction Generator Base Wind Turbine System," in 2018 International Conference on Electrical Sciences and Technologies in Maghreb (CISTEM), 28–31st October 2018, pp. 1–6.
7. Shi Wei, Xing Yan, Yundong Ma, Rixin Lai, "Modelling and Control for a Non-Grid-Connected Wind Power System Based on a Fixed-Pitch Variable Speed Wind Turbine and a Doubly Salient Generator," in 2009 World Non-Grid-Connected Wind Power and Energy Conference, 24–26th September 2009, pp. 1–4.
8. Farhan Ullah, Shahzad Ali, Deng Ying, Adnan Saeed, "Linear Active Disturbance Rejection Control Approach Base Pitch Angle Control of Variable Speed Wind Turbine," in 2019 IEEE 2nd International Conference on Electronics Technology (ICET), 10–13th May 2019, pp. 614–618.
9. Sriparna Das, Om Hari Gupta, "Study and Simulation of PMSG-Based Wind Turbine," in 2020 EPREC, NIT Jamshedpur, May 2020, pp. 37–46.
10. M. H. Zamani, G. H. Riahy, R. Z. Foroushani, "Introduction of a New Index for Evaluating the Effect of Wind Dynamics on the Power of Variable Speed Wind Turbines," in 2008 IEEE/PES Transmission and Distribution Conference and Exposition, 21–24th April 2008, pp. 1–6.
11. Hongwei Fang, Dan Wang, "A Novel Design Method of Permanent Magnet Synchronous Generator from Perspective of Permanent Magnet Material Saving," *IEEE Transactions on Energy Conversion*, vol.32, no-1, pp. 1–7, March 2017.
12. M. Kesraoui, N. Korichi, A. Belkadi, "Maximum Power Point Tracker of Wind Energy Conversion System," *Renewable Energy*, vol.36, no-10, pp. 2655–2662, October 2011.
13. Neeraj Keskar, Gabriel A. Rincon-Mora, "Designing an Accurate and Robust LC-Compliant Asynchronous Boost DC-DC Converter," in 2007 IEEE International Symposium on Circuits and Systems, 27–30th May 2007, pp. 549–552.
14. Arnab Ghosh, Subrata Banerjee, "Design of Type-III Controller for DC-DC Switch-Mode Boost Converter," in 2014, 6th IEEE Power India International Conference (PIICON), 5–7th December 2014, pp. 1–6.

15. Sachin Kumar Singh, Ahteshamul Haque, "Performance Evaluation of MPPT Using Boost Converters for Solar Photovoltaic System," in 2015 Annual IEEE India Conference (INDICON), 17–20th December 2015, pp. 1–6.
16. Saidi Khadidja, Maamoun Mountassar, Bounekhla M'hamed, "Comparative Study of Incremental Conductance and Perturb & Observe MPPT Methods for Photovoltaic System," in 2017 International Conference on Green Energy Conversion Systems (GECS), 23–25th March 2017, pp. 1–6.
17. Deepak Verma, Savita Nema, R. K. Nema, "Implementation of Perturb and Observe Method of Maximum Power Point Tracking in SIMSCAPE/MATLAB," in 2017 International Conference on Intelligent Sustainable Systems (ICISS), 7–8th December 2017, pp. 148–152.
18. D. Menniti, A. Burgio, N. Sorrentino, A. Pinnarelli, G. Brusco, "An Incremental Conductance Method with Variable Step Size for MPPT: Design and Implementation," in 2009 10th International Conference on Electric Power Quality and Utilisation, 15–17th September 2009, pp. 1–5.
19. M. Azzouzi, "Modeling and Simulation of a Photovoltaic Cell Considering Single-Diode Model," in Recent Advances in Environmental Science and Biomedicine, pp. 175–182.
20. Roberto F. Coelho, Filipe M. Concer, Denizar C. Martins, "A MPPT Approach Based on Temperature Measurements Applied in PV Systems," in IEEE ICSET, 2010, pp. 1–6.
21. Hong-sheng Chen, Tien-chien Jen, "A Passive Control Method of HAWT Blade Cyclical Aerodynamic Load Induced by Wind Shear," in 2017 International Conference on Mechanical and Intelligent Manufacturing Technologies (ICMIMT), 3–6th February 2017, pp. 106–109.
22. Yash Ajgaonkar, Mayuri Bhirud, Poornima Rao, "Design of Standalone Solar PV System Using MPPT Controller and Self-Cleaning Dual Axis Tracker," in 2019 5th International Conference on Advanced Computing & Communication Systems (ICACCS), 15–16th March 2019, pp. 27–32.
23. Siavash Taghipour Broujeni, Seyed Hamid Fathi, Javad Shokrollahi Moghani, "Hybrid PV/Wind Power System Control for Maximum Power Extraction and Output Voltage Regulation," in 3rd International Conference on Control, Instrumentation and Automation, 28–30th December 2013, pp. 59–64.
24. K. M. Muttaqi, M. T. Hagh, "A Synchronization Control Technique for Soft-Connection of Doubly-Fed Induction Generator-Based Wind Turbines to the Power Grid," in 2017 IEEE Industry Applications Society Annual Meeting, 1–5th October 2017, pp. 1–7.
25. Agata Dzionk, Robert Malkowski, "Activity Coordination of Capacitor Banks and Power Transformer Controllers in Order to Reduce Power Losses in the MV Grid," in 2016 10th International Conference on Compatibility, Power Electronics and Power Engineering (CPE-POWERENG), 29th June–1st July 2016, pp. 27–32.
26. Saroj Kumar Panda, Tusar Kumar Dash, "An Improved Method of Frequency Detection for Grid Synchronization of DG Systems During Grid Abnormalities," in 2014 International Conference on Circuits, Power and Computing Technologies [ICCPCT-2014], 20–21st March 2014, pp. 153–157.
27. Jinmok Lee, Jaeho Choi, "Decoupling Voltage Controller Design with Time Response Specifications for Three-Phase DC/AC Inverter," in 2008 IEEE 2nd International Power and Energy Conference, 1–3rd December 2008, pp. 630–634.
28. Liu Yang, Junrong Peng, Fan Yang, Yinghui Zhang, Hongfei Wu, "Single-Phase High Gain Bidirectional DC/AC Converter Based on High Step-Up Step-Down DC/DC Converter and Dual Input DC/AC Converter," in 2019 IEEE10th International Symposium on Power Electronics for Distributed Generation Systems (PEDG), 3–6th June 2019, pp. 554–559.

Renewable Energy Source Design in Microgrids

29. S. Rajanna, R. P. Saini, "Optimal Modelling of an Integrated Renewable Energy System with Battery Storage for Off Grid Electrification of Remote Rural Area," in 2016 IEEE 1st International Conference on Power Electronics, Intelligent Control and Energy Systems (ICPEICES), 4–6th July 2016, pp. 1–6.
30. S. Rajanna, R. P. Saini, "Optimal Modelling of Solar/Biogas/Biomass Based IRE System for a Remote Area Electrification," in 2014 6th IEEE Power India International Conference (PIICON), 5–7th December 2014, pp. 1–5.
31. Ajai Gupta, R. P. Saini, M. P. Sharma, "Optimized Application of Hybrid Renewable Energy System in Rural Electrification," in India International Conference on Power Electronics, 2006, pp. 337–340.
32. S. Rajanna, R. P. Saini, "GA Based Optimal Modeling of Integrated Renewable Energy System for Electrification of Remote Rural Area," in 2016 IEEE 6th International Conference on Power Systems (ICPS), 4–6th March 2016, pp. 1–6.
33. S. Pant, R. P. Saini, "Comparative Study of MPPT Techniques for Solar Photovoltaic System," in 2019 International Conference on Electrical, Electronics and Computer Engineering (UPCON), 8–10th November 2019, pp. 1–6.

4 Microgrid Power System Control Designs

Renuka Loka, Alivelu M. Parimi, and P. Shambhu Prasad

CONTENTS

4.1 Introduction 83
4.2 MG Power System Model..86
    4.2.1 Modeling of RTPS..86
    4.2.2 Modeling of MGES 87
        4.2.2.1 Modeling of WTG and PV..87
        4.2.2.2 Modeling of Diesel Engine Generator 88
        4.2.2.3 Modeling of FC and BESS 88
    4.2.3 Modeling of Inverter Delay, Filter Delay, and Communication Delay 88
4.3 MG Power System Control Architecture 89
    4.3.1 RTPS Control...89
    4.3.2 Distributed Primary Controller for FC and BESS............................92
    4.3.3 Distributed Primary Controller for Distributed Generation..............92
    4.3.4 Distributed Wind Speed-Dependent Frequency Controller for WTG..92
    4.3.5 Hierarchical Secondary Controller and Tertiary Controller for All MGES Other Than PV 93
4.4 Analysis Using Simulation Studies...93
    4.4.1 Wind Disturbance and Solar Irradiation Models 93
    4.4.2 Step Response under Different Parameter Disturbance Conditions...94
    4.4.3 Controller Performance under Random Disturbance Parameter Variations..97
4.5 Conclusion 98
References..99

4.1 INTRODUCTION

In microgrids, power can be shared among all micro sources depending on the frequency droop characteristic that can be used for primary frequency control of the MG, where droop can vary depending on the type of micro source [1]. Secondary control, an essential part of MG frequency control, can be achieved using a load

DOI: 10.1201/9781003121626-4

frequency control (LFC) loop by designing an LFC controller for MG sources [2]. For longer time scales, tertiary frequency control can be implemented as a third control loop to achieve economic optimization of MG power dispatch [3].

Microgrids suffer from various issues like low R/X ratios, high rate of rise of frequency, large swings in frequency caused due to improper power sharing and less inertia due to high renewable energy penetration which result in frequency stability problems, even within a few seconds after a small/large disturbance [4]. In order to overcome these issues, the design of hierarchical controls that are implemented at different time frames through primary, secondary and tertiary controls (PSTC) can be identified as an essential component for successful operation of MGs [5]. Tertiary and secondary control can be centrally operated, whereas distributed control can be implemented for primary frequency control [6]. Although time frames can be differentiated for PSTC, all these controls act together to perform frequency control. Thissignifies that the design of models that can account for various control actions provided using PSTC are necessary.

It can be seen from Figure 4.1 [7] that tertiary control sets the real power deviation command, ΔP_{ter}, every 15 min to 1 hr. Secondary control adds the real power deviation command, ΔP_{sec}, every 30 s to 15 min. Primary control adds the real power deviation command, ΔP_{pri}, from a few seconds to 5 min. The sum of these real power deviation commands participates in setting the controls needed to obtain the desired real power deviation for the MG.

In some LFC studies, simple first-order transfer function models were considered, neglecting inverter and communication delays [8]. On the other hand, communication delay modeling is important in developing detailed models [9]. Including communication delay models in frequency control models developed for an MG results in more realistic models that can help in designing more suitable LFC controllers.

Penetration of RE reduces the overall system inertia, which impacts frequency stability because of poor inertial response [10]. Virtual inertia (VI) for DFIGURE has been implemented to improve the system inertial response to obtain better frequency control [11]. However, VI for wind systems can impact system stability during the islanding condition of an MG. A recent study developed a decoupling method to suppress the negative effects of VI [12]. Although dynamic decoupling is achieved, peak deviation cannot be significantly reduced through this method.

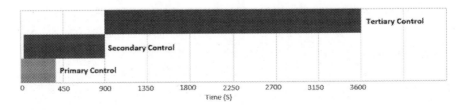

FIGURE 4.1 Hierarchical controls—timescales.

Moreover, it is important to consider wind speed variations and variations in solar isolation in order to address the uncertainties in power generation from RE. The MG power system control design should be insensitive to such variations. Step changes were considered to represent wind speed and solar isolation fluctuations in [8] and [13]. To mimic the real-time dynamics, more detailed models are necessary for robust frequency control design. An additional problem is restricting the MG model to a few energy sources [11–13], whereas the behavior of MG changes with the addition of another energy source.

The challenges faced by MG power system frequency control can be summarized as:

- Higher-order models for various energy sources considering the effect of communication delays to design robust PSTC should be given more importance.
- The uncertainties in wind speed or solar irradiation are not well represented to model real-time dynamic behavior of these energy sources.
- Hierarchical and distributed control architecture for MG models considering different MG sources incorporating PSTC has not been sufficiently addressed considering more detailed models.
- MG frequency control during both grid-connected mode and islanded mode should be satisfactory when VI is employed for wind turbine generators.

MG islanded mode of operation is widely accepted as complex in terms of frequency control because of high participation of RE with different droop characteristics. In view of the previously mentioned challenges, MG power system models and control architecture have been developed and simulated for frequency control.

This chapter mainly deals with the following aspects:

- Modeling the reheat thermal power system (RTPS) and MG to form an MG power system by including inverter and communication delays in order to design a robust frequency control in both operating modes.
- Establishing a PSTC for frequency control by considering hierarchical tertiary and secondary controls along with distributed primary control while including the detailed modeling of MG.
- Representing the RE uncertainties considering the real-time dynamic nature through developing suitable models for wind speed and solar irradiation.
- Developing a wind-speed-dependent frequency controller which operates smoothly during grid-connected mode, the transition phase and islanding mode of the MG power system even under wind speed uncertainties.

The remainder of this chapter is organized in the following manner. Section 4.2 deals with the detailed modeling of MG power systems, Section 4.3 describes the design and implementation of control architecture for frequency control of MG power system, Section 4.4 discusses the simulations results along with important discussion and Section 4.5 presents the conclusion.

4.2 MG POWER SYSTEM MODEL

MG power system components are depicted in Figure 4.2, where a reheat thermal power system is connected to the MG. Microgrid energy sources (MGES) are listed as follows:

- Wind turbine generator
- Fuel cell (FC)
- Battery energy storage system (BESS)
- PV generation
- Diesel generator

Modeling of RTPS and MGES is to be carried out in such a manner that the model clearly represents the changes in power deviation when subject to certain parameter variations. When the MG power system operates in grid-connected mode, RTPS participates in frequency regulation, and when the MG operates in islanded mode, the MGES alone participates in frequency regulation.

4.2.1 Modeling of RTPS

RTPS, inertia and load are modeled using the following equations:

$$\Delta f = \frac{1}{2HS + D} \Delta P_e \tag{4.1}$$

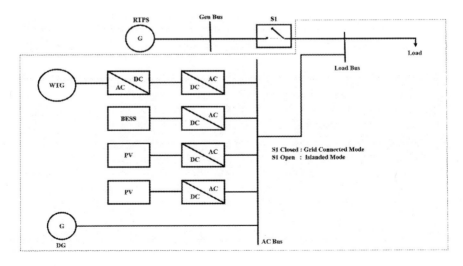

FIGURE 4.2 MG power system components.

Microgrid Power System Control Designs

$$\Delta P_e = \Delta P_{th} + \Delta P_{mg} - \Delta P_L \tag{4.2}$$

$$\Delta P_{th} = \frac{K_r T_r s + 1}{T_r s + 1} \Delta P_{t1} \tag{4.3}$$

$$\Delta P_{t1} = \frac{1}{T_t s + 1} \Delta P_t \tag{4.4}$$

$$\Delta P_t = \frac{K_g}{T_g s + 1} \Delta P_g \tag{4.5}$$

where 'ΔP_e', 'ΔP_{th}', 'ΔP_L' and 'ΔP_{mg}' are the real power fluctuations of the MG power system, RTPS, load demand and MG. 'ΔP_{t1}', 'ΔP_t' and 'ΔP_g' are the changes in command signals in a Stage II turbine, Stage I turbine and governor, respectively. 'Δf' is the frequency fluctuation. K_r and K_g are the gain constants, and T_r, T_t and T_g are the time constants of the turbine and governor models, respectively.

4.2.2 Modeling of MGES

MG power sources are modeled using the following equations:

$$\Delta P_{mg} = \Delta P_{wtg} + \Delta P_{dg} \pm \Delta P_{bess} + \Delta P_{fc} + \Delta P_{pv} \tag{4.6}$$

where 'ΔP_{dg}', 'ΔP_{bess}', and 'ΔP_{fc}' are the real power fluctuations in WTG, BESS and distributed generation, respectively.

4.2.2.1 Modeling of WTG and PV

The wind and PV equations are specified in the following [13]. The power generated by WTG is given by

$$P_{wt} = 0.5 \left(A_R C_p W_v^3 \right) \tag{4.7}$$

where β is air density, A_R is the turbine blade swept area, C_p is the power coefficient and W_v is the variation in wind velocity. The transfer function that represents the WTG is given as

$$G_{wtg}(s) = \frac{K_{wtg}}{T_{wtg} s + 1} \tag{4.8}$$

where $K_{wtg} = 1$ and $T_{wtg} = 1.5$ are the WTG's gain and time constants, respectively.
The power generated by PV generation is given by

$$P_{pv} = \eta S \varphi \{ 1 - 0.005 (T_a + 25) \} \tag{4.9}$$

where η, φ, and T_a represent conversion efficiency, irradiation in kW/m² and atmospheric temperature, respectively. The transfer function representation of PV is given by

$$G_{pv}(s) = \frac{K_{pv}}{T_{pv}s + 1} \tag{4.10}$$

where $K_{pv} = 1$ and $T_{pv} = 1.5$ are the PV's gain and time constants, respectively.

4.2.2.2 Modeling of Diesel Engine Generator

A diesel generator can be specified by a turbine-governor model [14]. The equations that represent the turbine-governor model are given as

$$G_{dg}(s) = \frac{1}{(T_{gdg}s + 1)(T_{tdg}s + 1)} \tag{4.11}$$

where $T_{gdg} = 0.08$ and $T_{tdg} = 0.3$ are the distributed generation's governor and turbine time constants, respectively.

4.2.2.3 Modeling of FC and BESS

BESS can either charge or discharge depending on the additional or deficit power generated by RE in the MG. The transfer function model that represents the BESS is given as [13]

$$G_{bess}(s) = \frac{K_{bess}}{T_{bess}s + 1} \tag{4.12}$$

where $K_{bess} = 0.03$ and $T_{bess} = 0.1$ are the BESS's gain and time constants, respectively.

FCs can generate power, but some power generated by FCs is consumed by the aqua-electrolyzer (AE), and the transfer function models for FC and AE are given as

$$G_{fc}(s) = \frac{K_{fc}}{T_{fc}s + 1} \tag{4.13}$$

$$G_{ae}(s) = \frac{K_{ae}}{T_{ae}s + 1} \tag{4.14}$$

where $K_{fc} = 0.01$ and $T_{fc} = 4$ are the BESS's gain and time constants, respectively, and $K_{ae} = 1$ and $T_{ae} = 0.08$ are the BESS's gain and time constants, respectively.

4.2.3 MODELING OF INVERTER DELAY, FILTER DELAY, AND COMMUNICATION DELAY

The inverter, filter and communication delays can be modeled as exponential functions [2] that can be represented in Laplacian form in the MG power system model. The transfer function models are given as

Microgrid Power System Control Designs

$$G_{inv}(s) = \frac{1}{T_{inv}s + 1} \quad (4.15)$$

$$G_{fil}(s) = \frac{1}{T_{fil}s + 1} \quad (4.16)$$

$$G_d(s) = \frac{1}{T_d s + 1} \quad (4.17)$$

where $T_{inv} = 0.04$, $T_{fil} = 0.004$ and $T_d = 0.1$ are the inverter, filter and delay time constants, respectively.

By considering various models, the complete block diagram representation of the MG power system model can be obtained as depicted in Figure 4.3, obtained by combining Equations (4.1)–(4.17).

4.3 MG POWER SYSTEM CONTROL ARCHITECTURE

An MG power system operates through hierarchical and distributed control architecture, as depicted in Figure 4.4. The controllers employed for the MG power system control are listed in the following:

- Primary controller for RTPS
- Hierarchical secondary controller for RTPS and tertiary control for RTPS
- Distributer primary controller for FC and BESS
- Distributed primary controller for distributed generation
- Distributed wind speed-dependent frequency controller for WTG
- Hierarchical secondary controller and tertiary controller for all MGES other than PV

Transport delay of the controller signal is modeled by using an equivalent transportation delay given by

$$Transport\ delay = e^{-sT_{td}} \quad (4.18)$$

4.3.1 RTPS Control

Primary control is designed on the basis of droop with a regulation constant of $R = 2.4 Hz/MW$. The equation for primary control signal is given as

$$u_1 = -\frac{\Delta f}{R} \quad (4.19)$$

where 'Δf' is the change in frequency. The LFC controller is designed using a proportional-integral-derivative (PID) controller, whose equation is given as

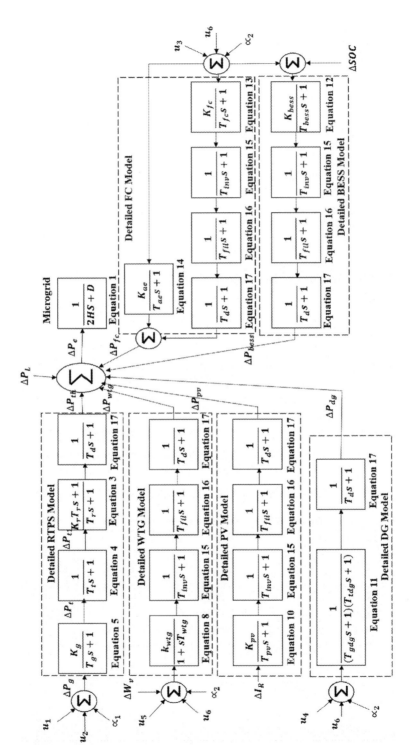

FIGURE 4.3 Detailed model of MG power system components.

Microgrid Power System Control Designs

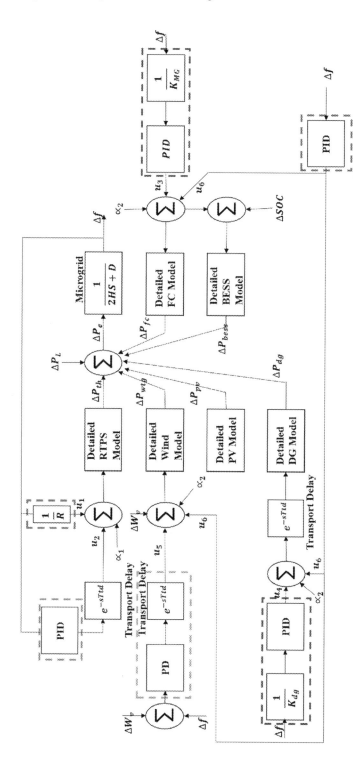

FIGURE 4.4 Hierarchical and distributed control architecture of MG power system.

$$u_2 = -\Delta f\left(k_{p1} + \frac{k_{i1}}{s} + k_{d1}s\right) \quad (4.20)$$

where k_{p1}, k_{i1} and k_{d1} are the controller constants. The tertiary control signal, α_1 is derived from the tertiary frequency control loop, which is a constant feedback loop.

4.3.2 Distributed Primary Controller for FC and BESS

A droop cascaded PID control is designed for the FC and BESS, whose control signal can be expressed by the equation

$$u_3 = \frac{-\Delta f}{K_{mg}}\left(k_{p2} + \frac{k_{i2}}{s} + k_{d2}s\right) \quad (4.21)$$

where $K_{mg} = 0.0083$ is the droop constant. k_{p2}, k_{i2} and k_{d2} are the controller constants.

4.3.3 Distributed Primary Controller for Distributed Generation

Based on the droop of the distributed generation governor, a droop-cascaded PID control is designed for primary control of distributed generation, and its equation is given as

$$u_4 = \frac{-\Delta f}{K_{dg}}\left(K_{p3} + \frac{k_{i3}}{s} + k_{d3}s\right) \quad (4.22)$$

where $K_{dg} = 0.05$ is the droop constant as considered for a Type-1 plant in [15]. k_{p2}, k_{i2} and k_{d2} are the controller constants.

4.3.4 Distributed Wind Speed-Dependent Frequency Controller for WTG

Wind speed is a randomly varying variable which causes large power swings in WTG generation. These large power swings result in frequency swings, which is undesirable. To control such power swings, a controller that controls the WTG generation based on the wind speed variation and frequency deviation is designed and implemented for distributed frequency control of the MG.

$$u_5 = (-\Delta f + \Delta W_v)(k_{pw} + k_{dw}s) \quad (4.23)$$

where k_{pw} and k_{dw} are the wind-speed controller constants.

A transport delay for this controller is also introduced to account for the effect of delay of the control signal.

4.3.5 HIERARCHICAL SECONDARY CONTROLLER AND TERTIARY CONTROLLER FOR ALL MGES OTHER THAN PV

An LFC control loop based on PID control is incorporated for MG sources other than PV, to achieve hierarchical control of a MG whose control equation is expressed as

$$u_6 = -\Delta f \left(k_{p4} + \frac{k_{i4}}{s} + k_{d4}s \right) \quad (4.24)$$

where k_{p4}, k_{i4} and k_{d4} are the controller constants. The tertiary control signal α_2 is derived from the tertiary frequency control loop, which is a constant feedback loop.

Distributed control architecture helps various energy sources participate in frequency control consistent with the load disturbances or RE disturbances. Equations (4.18)–(4.24) are combined along with detailed models of MG power system components, and the control architecture is as depicted in Figures 4.3 and 4.4 and utilized for simulation studies.

4.4 ANALYSIS USING SIMULATION STUDIES

The performance of various controls discussed in Section 4.3 when subject to parameter disturbance conditions while operating in different MG modes has been simulated to study the effectiveness of the proposed control architecture by including the detailed models discussed in Section 4.2. The response of the MG power system under different conditions can be analyzed through various simulations under different disturbance parameter conditions. The models discussed in Figures 4.3 and 4.4 are used for analysis in MATLAB/Simulink for dynamic time domain analysis. The disturbance parameter conditions for time domain analysis while performing simulation studies are listed in the following:

- A step load disturbance of 10%
- Wind speed disturbance of (8–11) m/s
- Solar irradiation levels increasing from 800 to 1000 w/m^2
- Solar irradiation levels decreasing from 800 to 400 w/m^2
- The disturbance parameter conditions are verified during two modes of MG operation:
  - MG power system in grid connected mode
  - MG power system in islanded mode

4.4.1 WIND DISTURBANCE AND SOLAR IRRADIATION MODELS

The disturbance model is shown in Figure 4.5. An interval and pulse generator block is used to simulate the time and amplitude of the wind speed or solar irradiation disturbance. A rate limiter is used to define the rate of rise and fall of the disturbance pulse to mimic the real-time dynamic changes of wind speed and solar irradiation.

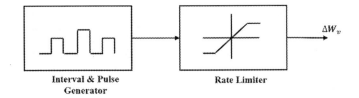

FIGURE 4.5 Wind disturbance model.

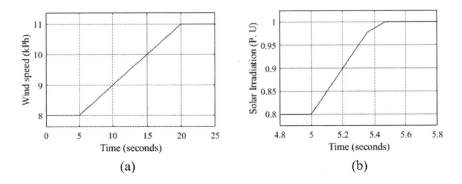

FIGURE 4.6 Deviations in (a) wind speed (b) solar irradiation.

The simulated changes in wind speed and solar irradiation are depicted in Figure 4.6. These models can represent the changes in a more appropriate way than considering a step disturbance.

4.4.2 Step Response under Different Parameter Disturbance Conditions

The performance of controllers should be such that they can respond to sudden and large step load changes so as to maintain frequency of the system during either grid-connected or islanded mode of operation. The response of controllers under various wind speed and solar irradiation changes should be suitable to maintain the frequency at nominal values.

To study the performance of controllers, various dynamic time-domain simulations are necessary to analyze the frequency deviation under different conditions. Simulation studies are presented by considering a wind speed, solar irradiation and step load disturbances and also by including a transport delay in the controller signal.

Initially, performance of distributed primary controllers is analyzed. For a 10% step load change at $t = 5s$ and wind-speed variation from 8 to 11 m/s at $t = 0$ s, with primary control acting alone on the MG power system, peak deviation is reduced by 25% when the solar irradiation is reduced from 0.8 to 0.4 P. U as depicted in Figure 4.7. It is evident that the amount of solar irradiation change can affect the transient frequency deviation of the MG power system.

Microgrid Power System Control Designs

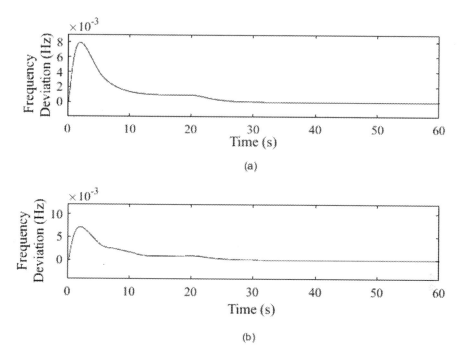

FIGURE 4.7 Step response of MG power system for (a) wind speed deviation from (8–11) m/s and solar irradiation deviation from 800 to 1000 w/m² (b) wind speed deviation from (8–11) m/s and solar irradiation deviation from 800 to 400 w/m².

As a next step, the performance of secondary and tertiary control during islanded mode is considered for analysis. It can be seen from the step response depicted in Figure 4.8 that when hierarchical control is implemented in the MG power system, along with distributed primary control, frequency oscillations are damped during islanded mode of operation considered at $t=50s$.

Peak overshoot in frequency deviation of the MG power system is mainly because of the changes in wind speed. To address this issue, a wind speed-dependent controller is employed, whose step response is depicted in Figure 4.8, which reduces the peak deviation by close to 90%, which proves the effectiveness of the designed wind speed-dependent controller. The transition to islanded mode is very smooth, with reduced peak overshoot during islanded operation when compared to distributed primary and hierarchical controls acting for frequency control.

The effect of transport delay on the wind-speed-dependent controller is a significant component to be studied. In Figure 4.9, a transport delay of 0.1s is considered for the controller to obtain the step response.

When the transport delay in wind-speed-dependent controller signal transmission is increased from 0.1 to 0.3s, the peak deviation in frequency is increased by 73%, as shown in Figure 4.10. It can be understood from this example that a transport delay in the control signal can significantly impact the controller performance

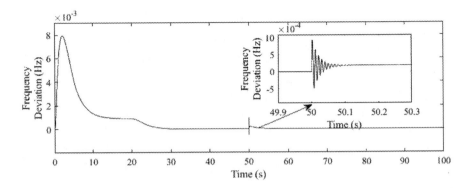

FIGURE 4.8 Step response considering secondary and tertiary control during islanded mode.

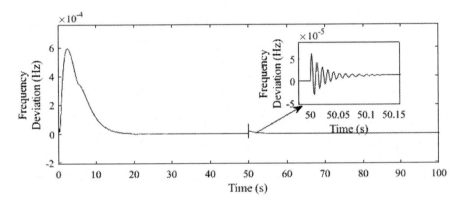

FIGURE 4.9 Step response of MG power system while implementing wind speed-dependent frequency controller.

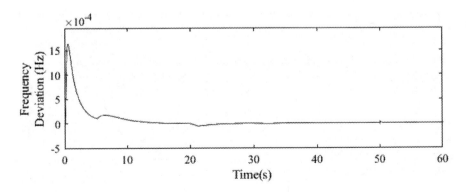

FIGURE 4.10 Effect of transport delay in wind-speed controller signal on step response of MG power system.

4.4.3 Controller Performance under Random Disturbance Parameter Variations

A wind profile varying between 7 and 11 m/s in a random manner is considered to analyze the controller performance. Random variations in wind speed can closely replicate the real-time wind patterns for which the effectiveness of the controller can be studied.

The random variations of a highly varying wind profile have been considered in the wind-speed model and wind-speed pattern, as shown in Figure 4.11(a). For the wind profile shown in Figure 4.11(a), the amount of wind power variation from WTG is in the range of ±0.2 P.U, as depicted in Figure 4.11(b); the frequency deviation is very minimal; and the frequency is close to nominal frequency, which clearly indicates the effectiveness of the wind-speed-dependent controller in achieving robust frequency control, which is shown in Figure 4.11(c).

The load variation is also a random phenomenon, and the design of control architecture should compensate for such load changes in order to maintain real power balance.

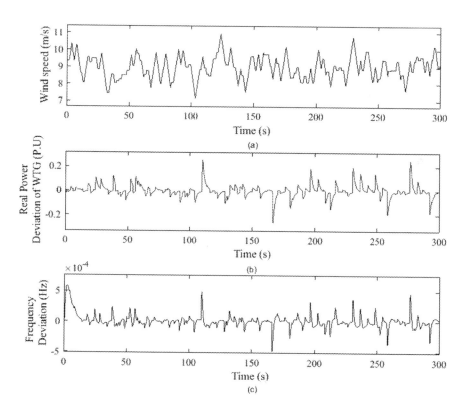

FIGURE 4.11 Deviation in (a) wind-speed (b) real power of WTG (c) frequency of MG power system.

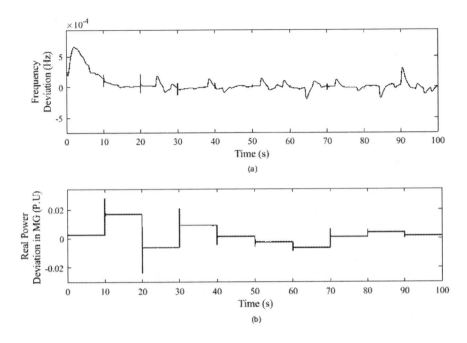

FIGURE 4.12 Deviation in (a) frequency (b) real power of MG power system.

Along with wind profile, a random load profile is also considered while islanding the MG power system at $t = 0$s, as depicted in Figure 4.12(a) and (b). The frequency deviation is minimal and the real power deviation of MG within the range of ±0.025 P.U.

The load variation considered dynamically varies with respect to time within a range of ±0.015 P.U. The response of controllers is quick and robust when time-varying load disturbance is considered.

4.5 CONCLUSION

Modeling of MG power systems considering inverter, communication and transport delays is essential for establishing more detailed models which can be utilized for designing a suitable control architecture. Depending on the droop of the MG energy source, distributed primary controls are designed, which are coordinated by hierarchically controlled secondary and tertiary controls. Wind speed uncertainties, which are the primary cause of large frequency swings, can be controlled by implementing a distributed wind speed-dependent controller. The designed control architecture is analyzed using various simulation studies. A 90% reduction in the peak frequency swing has been observed at a transport delay of 0.1 s. Random changes in load or wind speed in the case of islanded mode were efficiently handled by the MG power system control through limiting the power swings in the microgrid to a range of ±0.025 P.U.

REFERENCES

1. R. H. Lasseter, "MicroGrids," *2002 IEEE Power Engineering Society Winter Meeting. Conference Proceedings* (Cat. No.02CH37309), vol. 1 (New York, 2002): 305–308.
2. N. Vafamand, M. H. Khooban, T. Dragičević, et al., "Time-Delayed Stabilizing Secondary Load Frequency Control of Shipboard Microgrids," *IEEE Systems Journal*, vol. 13, no. 3 (Sept. 2019): 3233–3241, doi: 10.1109/JSYST.2019.2892528.
3. Z. Li, Z. Cheng, J. J. Liang, J. Si, L. Dong, et al., "Distributed Event-Triggered Secondary Control for Economic Dispatch and Frequency Restoration Control of Droop-Controlled AC Microgrids," *IEEE Transactions on Sustainable Energy* (2019), doi: 10.1109/TSTE.2019.2946740.
4. M. Farrokhabadi, et al., "Microgrid Stability Definitions, Analysis, and Examples," *IEEE Transactions on Power Systems*, vol. 35, no. 1 (Jan. 2020): 13–29.
5. B. Kirby, E. Ela, and M. Milligan, "Analyzing the Impact of Variable Energy Resources on Power System Reserves," *Renewable Energy Integration*, vol. 13, no. 6 (Elsevier, 2017): 85–101.
6. S. T. Cady, A. D. Domínguez-García, and C. N. Hadjicostis, "A Distributed Generation Control Architecture for Islanded AC Microgrids," *IEEE Trans. on Control Systems Technology*, vol. 23, no. 5 (Sept. 2015): 1717–1735, doi: 10.1109/TCST.2014.2381601.
7. P. Chilukuri, et al., "Introduction of Secondary Frequency Control in Indian Power System," *2018 20th National Power Systems Conference (NPSC)* (Tiruchirappalli, India, 2018): 1–6, doi: 10.1109/NPSC.2018.8771767.
8. T. Mahto, H. Malik, and M. Saad Bin Arif, "Load Frequency Control of a Solar-Diesel Based Isolated Hybrid Power System by Fractional Order Control Using Partial Swarm Optimization," *Journal of Intelligent & Fuzzy Systems (JIFS)*, vol. 35 (2018): 5055–5061.
9. S. Liu, X. Wang, and P. X. Liu, "Impact of Communication Delays on Secondary Frequency Control in an Islanded Microgrid," *IEEE Trans. on Industrial Electronics*, vol. 62, no. 4 (Apr. 2015): 2021–2031, doi: 10.1109/TIE.2014.2367456.
10. H. Bevrani, B. Francois, and T. Ise, *Microgrid Dynamics and Control* (New York: Wiley, 2017).
11. X. Tian, W. Wang, Y. Chi, Y. Li, and C. Liu, "Virtual Inertia Optimisation Control of DFIGURE and Assessment of Equivalent Inertia Time Constant of Power Grid," *IET Renewable Power Generation*, vol. 12, no. 15 (19 Nov. 2018): 1733–1740, doi: 10.1049/iet-rpg.2018.5063.
12. J. Xi, H. Geng, and X. Zou, "Decoupling Scheme for Virtual Synchronous Generator Controlled Wind Farms Participating in Inertial Response," *Journal of Modern Power Systems and Clean Energy* (2019) doi: 10.35833/MPCE.2019.000341.
13. K. Singh, M. Amir, F. Ahmad and M. A. Khan, "An Integral Tilt Derivative Control Strategy for Frequency Control in Multimicrogrid System," in *IEEE Systems Journal*, (2020), doi: 10.1109/JSYST.2020.2991634.
14. Y. Mi, Y. Song, Y. Fu, X. Su, C. Wang, and J. Wang, "Frequency and Voltage Coordinated Control for Isolated Wind–Diesel Power System Based on Adaptive Sliding Mode and Disturbance Observer," *IEEE Trans. on Sustainable Energy*, vol. 10, no. 4 (Oct. 2019): 2075–2083, doi: 10.1109/TSTE.2018.2878470.
15. C. Wu, et al., "Studies on Operation Modes of Diesel Generators in a Standalone Power System," *2018 4th International Conference on Green Technology and Sustainable Development (GTSD)* (Ho Chi Minh City, 2018): 72–76, doi: 10.1109/GTSD.2018.8595556.

5 Hybrid Microgrid Design Based on Environment, Reliability, and Economic Aspects

Sriparna Roy Ghatak, Aashish Kumar Bohre, and Parimal Acharjee

CONTENTS

5.1	Introduction	102
5.2	Uncertainty Modeling	103
	5.2.1 Probabilistic Model of Solar Irradiance	103
	5.2.2 Probabilistic Modeling of Wind Speed	104
	5.2.3 Load Forecasting	105
5.3	Battery Storage Model	105
5.4	Scenario Modeling Using Data Clustering	106
5.5	Modeling of Fitness Function	106
	5.5.1 Reliability Assessment of Microgrids	106
	5.5.2 Economic Aspects	106
	5.5.3 Environmental Aspects	107
5.6	Security Constraints	108
	5.6.1 Voltage Constraints	108
	5.6.2 Power Flow Constraints	108
	5.6.3 Power Equality Constraints	108
5.7	Design Problem Formulation	108
5.8	Multi-Objective Algorithm	109
	5.8.1 Theory of Parental Inheritance	109
	5.8.2 Non-Dominant Sorting	109
	5.8.3 Crowding Distance	109
	5.8.4 Dynamic Crowding Distance	109
	5.8.5 Compromise Solution	110
5.9	Case Study	110
5.10	Conclusion	114
References		117

DOI: 10.1201/9781003121626-5

5.1 INTRODUCTION

In recent years, due to high climatic and environmental concerns, countries across the globe are vastly shifting towards renewable and sustainable sources to meet their increasing energy demands. In this context, the concept of the microgrid, which consists of distributed energy sources, storage devices and controllable loads [1], is gaining huge importance worldwide. Microgrids are regarded as a critical technology which can enhance the resilience of power systems. They can operate either in grid-connected mode or independent or islanded mode [2]. A microgrid consists of the integration of various types of distributed generation units, such as a solar energy source, wind, fuel cells and so on, and interconnected loads and acts as single controllable entity [3]. Renewable energy sources such as solar and wind are uncertain or intermittent and mostly depend on additional aspects such as climatic and meteorological conditions and time of day [4]. In these circumstances, a hybrid system consisting of a combination of renewable sources such as solar and wind, along with storage in a microgrid environment, is an effective means to reduce environment pollution and meet energy demands. System uncertainty mainly arises due to intermittent renewable energy resources such as solar and wind. Solar and wind power is random in nature since its power output depends on atmospheric conditions such as solar irradiance, temperature, wind speed and so on [5-6]. A hybrid system consisting of two or more distributed energy sources can make the microgrid resilient to weather changes or fuel supply chain disruptions [7-8]. Energy storage devices are an important component of microgrids. They store the excess energy renewable energy when there is a surplus and dispatch it appropriately when needed [9-10]. For successful implementation and gradual commercialization of renewable-based microgrid systems, it is essential to develop a robust design framework considering a high level of system uncertainty and reliability. The optimal design of the system includes identifying the most suitable configuration and selecting the optimal capacities of the storage devices and energy sources [11].

Designing a microgrid is a complicated task, as it needs development of accurate mathematical models intended for each system component [12]. Capacity optimization of renewable sources is an important aspect of robust system design, which refers to determination of the optimal number of components such that system reliability is maximized and total cost of the system is minimized [13]. Oversizing a component may ensure high system reliability, but the economic benefit will be reduced. The unit sizing of generating sources has a huge impact on system efficiency [13]. Effective energy management and control is critical for proper microgrid operation. The primary aim of any efficient energy management strategy is to provide a stable supply of energy to consumers. A smart energy management system receives real-time data from the system, processes it and makes the appropriate control decision quickly [14]. The optimal design problem of microgrids is highly complex, nonlinear and nonconvex in nature and requires an efficient and robust optimization algorithm [13]. As multiple objectives are considered for the optimal design problem of microgrids, instead of conventional heuristic process in which various objectives are converted to a single objective problem, an effective multi-objective algorithm should be used [15-16].

Hybrid Microgrid Design

In this work, an uncertainty modeling framework of renewable sources such as solar and wind is interfaced with a multi-objective optimization scheme for the development of a viable design tool for a microgrid system. A battery system is considered as the storage system available in the proposed design model. Effective energy management strategies are proposed to make a secure and reliable supply of energy to load. It would be computationally expensive to consider all the uncertain scenarios of renewable energy and load while designing the microgrid. On the other hand, if all these scenarios are not considered, then the design model of the microgrid will not represent its true stochastic characteristic. Therefore, clustering techniques such as k-means clustering [16] are used in this work to lower the number of scenarios to a representative number that reflect the true stochastic nature of the microgrid. The design problem is resolved considering maximization of economic profit, reliability and environmental aspects. Security constraints of voltage, power flow and power equality are also considered in the proposed design strategy. A multi-objective algorithm, the extended non-dominant sorted genetic algorithm (E_NSGAII) [16], is used in this work to find the optimal capacity and location of solar, wind and battery storage systems considering the maximization of economic benefits, environment, and reliability. The proposed design model is tested in an IEEE 13 bus system [17] which is considered a microgrid for validation of the proposed design framework.

5.2 UNCERTAINTY MODELING

Modeling of intermittent solar energy sources and wind power is an important aspect of microgrid design [5, 16]. The modeling of hybrid energy sources consisting of solar, wind and a storage system is given as follows.

5.2.1 Probabilistic Model of Solar Irradiance

Currently, solar energy source has emerged as an appropriate choice for a sustainable energy supply in microgrids. The output of solar energy sources is highly uncertain in nature and varies with time, weather conditions, solar irradiance and temperature. This uncertainty or intermittent characteristics of solar energy sources creates challenges for system operators in balancing generation with load demand and maintaining system constraints. Therefore, in order to design a safe and reliable microgrid, it is essential to model the solar energy source considering its randomness. To express the random solar output, a beta probability density function $f(s^t)$ [16] is used.

$$f(s^t) = \begin{cases} \dfrac{\Gamma(a+b)}{\Gamma(a)\Gamma(b)}(s^t)^{a-1}(1-s^t)^{b-1} & \text{if } 0 \leq s^t \leq 1\, a,b > 0 \\ 0 & \text{otherwise} \end{cases} \quad (5.1)$$

The shape parameters are represented as a and b and can be expressed as follows

$$a = \frac{\mu \times b}{1-\mu} \qquad b = (1-\mu) \times ((\mu(1+\mu))/\sigma^2 - 1) \quad (5.2)$$

In Equation (5.2), σ and μ represent the standard deviation and mean, which are evaluated from historical data. The maximum power $P(s)$ of a PV system is formulated as follows.

$$P(s) = N \times FF \times V \times I \tag{5.3}$$

N in Equation (5.3) represents the module number of PV panels. FF is the fill factor, and V and I are the voltage and current of the solar array as given in Equations (5.4) and (5.5).

$$V = V_{Oc} - K_{Voltage} \times T_{Cell} \tag{5.4}$$

$$I = s^t[I_S + K_I \times (T_{Cell} - 25)] \tag{5.5}$$

V_{Oc} and I_S in Equations (5.4) and (5.5) are the voltage on an open circuit and current on a short circuit, respectively. T_{Cell}, K_I and $K_{Voltage}$ are the cell temperature of the solar panels and coefficient of temperature of current and voltage, respectively. To enhance the accuracy of the outcome, every hour is segregated into a number of states (NS). The probability of all the states is evaluated as given in Equation (5.1). The expected hourly output of solar power is calculated as follows:

$$P(s)_{hour} = \int_{i=1}^{NS} f(s^t) \times P_o(s^t) \tag{5.6}$$

5.2.2 Probabilistic Modeling of Wind Speed

Wind energy systems are another promising renewable energy source used in microgrids. The output of wind energy is variable or uncertain, as it is dependent on wind speed, which in turn depends on weather conditions, geographical location and so on. Therefore, it is essential to model the wind energy resource considering the system uncertainty. To express the randomness in wind speed, a Weibull probability distribution function [18–19] can be used as expressed in the following.

$$f_s(w) = \frac{K}{C}\left(\frac{w}{C}\right)^{K-1} exp\left(-\left(\frac{w}{C}\right)^K\right) \tag{5.7}$$

$$K = \left(\frac{\sigma}{\mu}\right)^{-1.086} ; C = \frac{\mu}{\Gamma\left(1 + 1/k\right)} \tag{5.8}$$

w, K and C in equation (5.7) represent wind speed and parameters representing shape and scale, respectively. σ and μ represent the standard deviation and mean of wind speed, which can be collected from historic data.

$$P_{wt}^t = \sum_{s=1}^{N_s} P_{wt}^s \times p_w(s) \tag{5.9}$$

Hybrid Microgrid Design

$$p_w(s) = \int_{w_{sl}}^{w_{su}} f_s(w) dw \qquad (5.10)$$

P_{wt}^s and $p_w(s)$ are the wind power output and the Weibull probability of wind speed, respectively, in the states. w_{su} and w_{sl} are the upper and lower limits of wind speed.

5.2.3 Load Forecasting

Power demand in microgrids exhibits distinctive uncertainty, as it varies with time, season and so on. To model the random nature of real and reactive loads, a probabilistic method of the Gaussian normal distribution ($f(l_t)$) [16] can be used, which is given as follows.

$$f(l_t) = \frac{1}{\sigma\sqrt{2\pi}} e^{\frac{-(l_t-\mu)^2}{2\sigma^2}} \qquad (5.11)$$

σ and μ in Equation (5.9) represent the standard deviation and mean of historic load data.

$$P(t) = P_{o(t)}(1+growth)^k + f(l_t)\sigma \qquad (5.12)$$

$$Q(t) = Q_{o(t)}(1+growth)^k + f(l_t)\sigma \qquad (5.13)$$

$P(t)$ and $Q(t)$ are the expected real and reactive power at time t. $P_{o(t)}$ and $Q_{o(t)}$ are the real and reactive power measured with a smart meter at time t. In Equations (5.10) and (5.11), growth represents load growth in the future and is considered in the process of designing the microgrid.

5.3 BATTERY STORAGE MODEL

Storage devices such as batteries are an extremely important aspect of microgrids. Battery storage devices are used to store the excess power of renewables and can be suitably used afterwards when needed in the system. In recent years lithium-ion batteries [9] have become a widely accepted storage solution for microgrids. The mathematical equation of energy storage during charging and discharging of battery storage $W(t)_B$ at a given time t is given as follows.

$$W(t)_B = W(t-1)_B + \eta_{charging} \times P_{ch}\Delta t \qquad (5.14)$$

$$W(t)_B = W(t-1)_B - \frac{P_{dis}\Delta t}{\eta_{discharging}} \qquad (5.15)$$

P_{ch} and P_{dis} are charging and discharging power, respectively. $\eta_{discharging}$ and $\eta_{charging}$ are the discharging and charging efficiency of the battery storage system, respectively.

The battery state of charge (SOC) is an important parameter for energy management of microgrids consisting of hybrid energy sources. For safe and reliable operation of the microgrid battery, the SOC should be within permissible limits.

5.4 SCENARIO MODELING USING DATA CLUSTERING

System uncertainty is an integral part of microgrids because of the high penetration of intermittent renewable energy sources in it. Additionally, system load is also uncertain and varies with time, consumer activities and so on. A group of clusters is made in this method in such a manner that the mean square distance from every data point of the original set and the representative data point of the created cluster known as the centroid is the minimum, as given in Equation (5.16).

$$J = \sqrt{\sum_{i=1}^{k} \sum_{x_i}^{n} (x_i - c_k)^2} \tag{5.16}$$

where c_k represents the centroid of the cluster.

5.5 MODELING OF FITNESS FUNCTION

The detailed modeling of the fitness function is given as follows.

5.5.1 Reliability Assessment of Microgrids

The reliability of microgrids is an important aspect which should be considered for system design. A reliable hybrid microgrid system should have the capability to supply consistent power to the load for a given period. Expected energy not served (EENS) can be used as the reliability index [16] in the microgrid and is precisely expressed as follows

$$EENS = \sum_{i=1}^{m} U_i \times P_i \tag{5.17}$$

where U_i represents unavailability and P_i represents real power load.

5.5.2 Economic Aspects

The unit cost of component $i(C_i)$ includes installation cost, replacement cost, salvage cost, operation cost and maintenance cost. In a microgrid, the components that are used are PV panels, wind system, battery storage converters and inverters.

$$C_i = \{Initial_{cost\,i} * P + R_i * CCP * P + O\&MC * CCP * YR_{span} * P - S\} \tag{5.18}$$

where $Initial_{cost\,i}$, $O\&MC$ R_i, S represents preliminary installation cost/KW, operation and maintenance cost, replacement cost and cost related to salvage, respectively. Cumulative current price (CCP) is allied with the variable cost of such operation and

Hybrid Microgrid Design

maintenance. *CCP* as termed in [16] is used to represent the complete costs during the given span of time. *CCP* is featured for expected future cash flows considering the rate of inflation and interest. Considering the inflation rate (*IFL*) and interest rate (*INR*), the current price (*CP*) is estimated as follows:

$$CP = \frac{1 + IFL/100}{1 + INR/100} \tag{5.19}$$

From *CP*, *CCP* is estimated for microgrid design as follows.

$$CCP = \frac{1 - CP^{YR_{span}}/100}{1 - CP} \tag{5.20}$$

$$S_i = R_i \frac{RL}{YR_{span}} \tag{5.21}$$

where S_i is the salvage cost of the component i and *RL* is the remaining life. The total cost $Total_{cost}$ incurred is mathematically expressed as follows.

$$Total_{cost} = N_{PV}C_{pv} + N_{wind}C_{wind} + N_{battery}C_{battery} + N_{inverter}C_{inverter} + C_{EENS} \tag{5.22}$$

where N_{PV}, N_{wind}, $N_{battery}$ and $N_{inverter}$ refer to number of PV panels, wind turbine, battery and inverter. C_{EENS} is the cost of energy not served. The revenue earned is from selling electric energy to consumers.

$$Total_{benefit} = \sum_{sc=1}^{N} P_{sc} C_{tariff} \times CCP \tag{5.23}$$

The benefit cost ratio (BCR) of microgrid design can be calculated as

$$BCR = \frac{Benefit}{Cost} \tag{5.24}$$

From Equation (5.24), it is obvious that a high *BCR* implies high economic benefits.

5.5.3 Environmental Aspects

To reduce pollution and encourage the usage of renewables, countries across the world are imposing carbon discharge rates on consumers and producers. For microgrids, to judge the environmental benefits due to integration of renewables, the environment benefit index (EBI) is used. The mathematical expression of EBI is given as follows.

$$EBI = \frac{(CE)_{without\_renewable} - (CE)_{with\_renewable}}{(CE)_{without\_renewable}} \tag{5.25}$$

where $(CE)_{without\_renewable}$ and $(CE)_{with\_renewable}$ are the cost of environmental emissions without and with the use of renewables in the microgrid, respectively. The higher the EBI ratio, higher the environmental benefits due to incorporation of renewables in the microgrid.

5.6 SECURITY CONSTRAINTS

Abiding by security constraints is the most important aspect for safe, secure operation of microgrids. Therefore, security constraints must be considered while designing the microgrid.

Some of the security constraints that should be considered are as follows.

5.6.1 Voltage Constraints

In microgrids, the voltage profile of the system should be within the upper and lower limits as given in Equation (5.26).

$$\left|\overline{V_i}\right|^{min} \leq \left|\overline{V_i}\right| \leq \left|\overline{V_i}\right|^{max} \tag{5.26}$$

5.6.2 Power Flow Constraints

Power flow should also be within limits, as given in equation (5.27).

$$S_i \leq S_i^{max} \tag{5.27}$$

5.6.3 Power Equality Constraints

At every time instant t, the total power generation should be equal to total load, as given in equation (5.28).

$$P_{grid,t} + P_{renewable,t} - P_{load,t} = 0 \tag{5.28}$$

5.7 DESIGN PROBLEM FORMULATION

In this chapter, a design problem of microgrids is resolved considering the maximization of economy, ecological benefits and reliability while considering security constraints as given in Section 5.3. Financial and ecological parameters such as cost of renewables, power tariff, grid power cost, inflation rate and interest rate are considered as the input to the problem. The reliability index, EENS, is already integrated in BCR and therefore not distinctly taken as a separate objective function. The design strategy consists of decisions concerning the number of devices to be used to maximize the objectives. To enhance the accuracy of the obtained results, the uncertainty of renewables and load is considered in the proposed problem.

5.8 MULTI-OBJECTIVE ALGORITHM

In microgrid design, the objectives to be fulfilled may be contradictory in nature. Therefore, it is essential to utilize a robust multi-objective algorithm to solve it. The extended non-dominant sorted genetic algorithm is utilized in the current process to solve the proposed design problem. Various operators of E_NSGAII are as follows.

5.8.1 Theory of Parental Inheritance

An improved pool of parents improves the convergence speed to the global Pareto front. Therefore, in E_NSGAII, chromosomes with the best results in terms of objective functions are selected and copied to the parent population. The process of creating the improved pool continues until the chosen size of the population is reached. If the problem is a single objective of population size N, it should be generated N times to create the pool. Along similar lines, if it is a multi-objective problem of P objectives, then number of generations will be N/P. Therefore, it is evident that the time for computation reduces with an increase in the number of fitness functions.

5.8.2 Non-Dominant Sorting

It is a process to rank solutions. The non-dominant solution in a pool is the solution that is not dominated by any other solution in terms of the fitness function. Every solution of the first front is given a rank of 1. These first front solutions are momentarily ignored and the next set of non-dominant solutions is found in the population. They are assigned a rank of 2 and are members of the second front. This process continues and stops when every member of the population pool is allocated a rank.

5.8.3 Crowding Distance

Crowding distance (d_i) is required to provide an estimate of the solution density adjacent to a particular solution i in the pool of the population. Solutions with high d_i are preferred, as they are placed in a low crowding zone. Thus, solutions with a large crowding distance are preferred to maintain diversity.

5.8.4 Dynamic Crowding Distance

To maintain diversity, the concept of dynamic crowding distance (D_i), as proposed in [6], is used in which the calculations for individuals are done adaptively. D_i is calculated in the non-dominant set for every solution, and among them, the solutions having the lowest D_i are discarded. Subsequently D_i is recalculated for the rest of the solutions, and the process continues until the requisite number of solutions is selected from the set of non-dominant solutions. The mathematical expression of D_i is as follows.

$$D_i = \frac{d_i}{\ln \frac{1}{V_i}} \qquad (5.29)$$

5.8.5 COMPROMISE SOLUTION

To select a compromise solution out of all the acquired Pareto solution set, fuzzy logic [20] can be used. The compromise solution should be at a high distance from the worst solutions and a small distance from the ideal solution. The mathematically fuzzy membership function of the *j*th solution is expressed as follows.

$$\mu_i^j = \frac{OF_i^{max} - OF_i}{OF_i^{max} - OF_i^{min}} \tag{5.30}$$

In equation (30), OF represents the objective functions. As the proposed design problem is a maximization problem, the maximum value of the objective function (OF_i^{max}) represents the best solution, and the minimum value of the objective function OF_i^{min} represents the worse solution.

5.9 CASE STUDY

In the present design problem, an IEEE 13 bus test system is considered as a microgrid. This system consists of two capacitors; one is connected at bus number 675 and another at bus 611. The microgrid system also consists of a voltage regulator placed between 650 and 632.

A diagram of the test microgrid is shown in Figure 5.1.

The E_NSGAII algorithm, as in [6], is used to resolve the problem in this work. Parent inheritance, non-dominant sort and dynamic crowding distance procedures are used in the multi-objective optimization. For the process of selection, the tournament

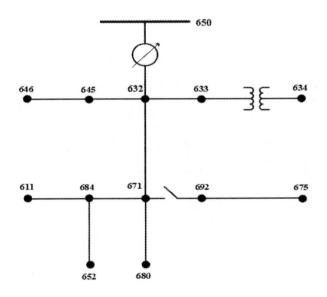

FIGURE 5.1 Single line diagram of IEEE 13 bus test system.

Hybrid Microgrid Design

selection method is used. The simulated binary crossover and polynomial mutation process are used for the crossover and mutation process in the E_NSGAII algorithm, respectively. Fuzzy logic is used to obtain the single compromise solution among the obtained Pareto solutions. Table 5.1 shows the parameters utilized in the multi-objective E_NSGAII process. The present design problem is framed considering a time span of 15 years. Table 5.2 shows the environment and economic data used in this work.

Figure 5.2 shows the representative scenarios considered in a year using k-means clustering. The historic data of solar, wind and load of four seasons (summer, winter, spring and autumn) are collected. The collected data is then divided into night and

TABLE 5.1
Parameters Used in E_NSGAII

Parameters	Values
Iteration number	450
Population size	100
Crossover Probability	0.9
Mutation Probability	0.5
Crossover Index	3
Mutation Index	18

TABLE 5.2
Economic and Environment Data

Parameters	Values
CE_{CO2}	10 $/ton CO_2
Inflation rate	10%
Interest rate	10%
PV	
Capital cost and replacement cost	1084$/KW
Operation and maintenance (O&M) cost	5$/year
Lifetime	20 years
Wind	
Capital cost and replacement cost	600$/KW
O&M cost	0.02$/KWh
Lifetime	20 years
Inverter	
Capital cost and replacement cost	127$/KW
Battery	
Capital and replacement cost	600$/KWh
O&M cost	1$/year
Lifetime	8years

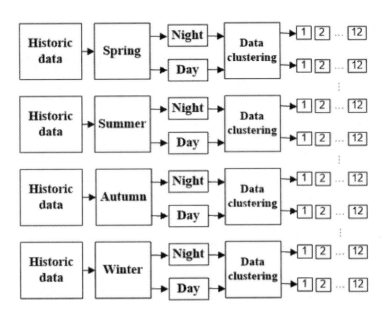

FIGURE 5.2 Representative scenarios.

day blocks. The value of K is chosen as 12, data clustering is used in every block and 96 representative scenarios are obtained in a year.

In Table 5.3, an energy management strategy of a hybrid energy source with an energy storage device is shown. The 24 hours in a day are divided into two time spans, 6 a.m. to 6p.m. and then 6 p.m. to 6 a.m. Each time period is divided into two power conditions.

Case I: When total generation is more than demand.
Case II: When total demand is more than generation.

Battery SOC is the most important control parameter, and accordingly switching decisions are made. For each case, five conditions of battery SOC are considered. The symbol ☑ indicates that the device is on, and the symbol ☒ indicates that the device is off in Table 5.3. Similarly, the symbol ↑↓ represents the charging status and discharging of the battery. Battery SOC is maintained between 40 and 90, and whenever the battery SOC reaches less than 40, the battery is charged. In this work grid, the connection is mostly switched off to ensure autonomous operation of the system. As shown in Table 5.3, the grid is switched on only during the night when both power from the battery and wind is insufficient to supply the load. At every point in time, safe and reliable power supply to the load will be ensured. Figure 5.3 shows the flowchart for the implementation of the E_NSGAII algorithm.

Figures 5.4 and Figure 5.5 show the expected hourly variation of solar output and wind output obtained using the beta probability density function and Weibull function as given in Equations (5.1) and (5.7), respectively. As is observed in Figure 5.4, the solar output is maximum at 1p.m. and reduced to 0 from 7p.m. onwards. In a

TABLE 5.3
Energy Management Strategy of the Hybrid Energy Source System

Time	Power Condition	PV Status	Battery SOC	Wind Status	Battery Status	Grid
6 a.m.–6p.m.	$P_{gen} \leq P_{load}$	☑	90–80	☒	↓	☒
		☑	80–60	☒	↓	☒
		☑	60–80	☒	↓	☒
		☑	40–60	☑	↓	☒
		☑	Less than 40	☑	↑	☒
	$P_{load} \leq P_{gen}$	☑	90–80	☒	↑	☒
		☑	80–60	☒	↑	☒
		☑	60–80	☒	↑	☒
		☑	40–60	☒	↑	☒
		☑	Less than 40	☑	↑	☒
6 p.m.–6a.m.	$P_{gen} \leq P_{load}$	☒	90–80	☑	↓	
		☒	80–60	☑	↓	☑
		☒	60–80	☑	↓	☑
		☒	40–60	☑	↓	☑
		☒	Less than 40	☑	↑	☑
	$P_{load} \leq P_{gen}$	☒	90–80	☑	↓	☒
		☒	80–60	☑	↓	☒
		☒	60–80	☑	↑	☒
		☒	40–60	☑	↑	☑
		☒	Less than 40	☑	↑	☑

similar way, the wind power output varies throughout the day and totally depends on wind speed.

Table 5.4 refers to the optimal location and size allocated for hybrid energy sources and battery energy storage. The energy sources and storage devices are allocated at the same location, as observed from Table 5.4. In Table 5.5, the objective function values, BCR and EBI, of the five best solutions are shown. As given in Section 5.8, the Pareto solutions with the highest value of dynamic crowding distance are the best solutions. However, solution 3 is the best one with respect to EBI and solution 4 for BCR. Therefore, it becomes essential to find one compromise solution among the best acquired solutions. The compromise solution is selected by using fuzzy logic, as in Section 5.5. Solution 2 is selected in this work as the compromise one and is marked bold in Table 5.5. It is clearly observed from the objective function values obtained in Table 5.5 that the optimum allocation of hybrid renewable sources and storage device provides economic and environment benefits. Figure 5.6 shows the

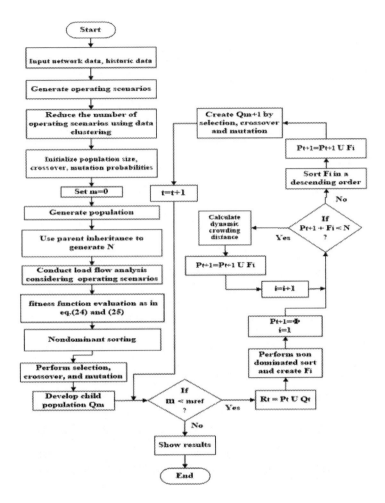

FIGURE 5.3 Flowchart.

optimal Pareto curve obtained through applying the multi-objective optimization procedure E_NSGAII. The figure shows all the obtained non-dominant curves using the multi-objective E_NSGAII procedure.

5.10 CONCLUSION

A robust design framework of a microgrid consisting of hybrid energy sources and battery storage is proposed in this chapter. The present design model aims to maximize financial, reliability and environmental benefits while considering the security constraints of the microgrid. Practical features such as renewables and load uncertainty are considered in this design structure. To reduce the complexity arising due

Hybrid Microgrid Design

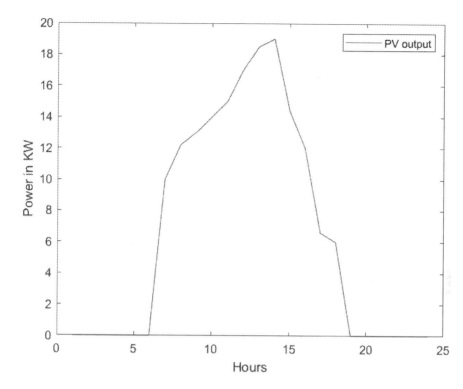

FIGURE 5.4 Hourly variation of solar power.

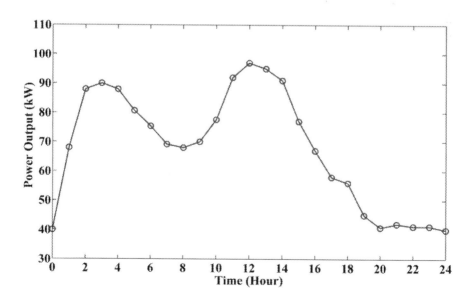

FIGURE 5.5 Hourly variation of wind power output.

TABLE 5.4
Optimal Location and Configuration of PV, Wind and Battery Storage

Location	Number of PV	Number of Wind	Number of Battery
671	20	10	5
675	25	15	10
652	15	10	5

TABLE 5.5
Objective Function Value of Best Pareto Solutions

Solutions	EBI	BCR
1	0.8112	4.824
2	**0.8134**	**4.162**
3	0.8312	5.271
4	0.7840	5.822
5	0.7520	2.811

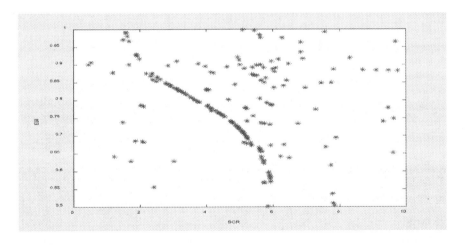

FIGURE 5.6 Pareto optimal curve.

to a large number of operating scenarios, k-means data clustering is utilized in this chapter. The fuzzy logic-based multi-objective E_NSGAII is successfully applied to find the optimal location and configuration of the renewable energy source and storage system. An IEEE 13 bus system is used as the test case for implementation of the design model. The results clearly indicate that the implementation of the present design methodology enhances the financial, reliability and ecological benefits of the system.

REFERENCES

1. P. Yang, and A. Nehorai. "Joint optimization of hybrid energy storage and generation capacity with renewable energy." *IEEE Transactions on Smart Grid* 5 no. 4 (2014): 1566–1574.
2. J. Nelson, N. G. Johnson, K. Fahy, and T. A. Hansen. "Statistical development of microgrid resilience during islanding operations." *Applied Energy* 279(2020): 115724.
3. S. Newman, K. Shiozawa, J. Barrett, E. Douville, T. Hardy, and A. Solana. "A comparison of PV resource modeling for sizing microgrid components." *Renewable Energy* 162 (2020):831–843.
4. P. Gangwar, S.N. Singh, and S. Chakrabarti. "Multi-objective planning model for multiphase distribution system under uncertainty considering reconfiguration." *IET Renewable Power Generation* 13, no. 12 (2019): 2070–2083.
5. M. Fan, V. Vittal, G. T. Heydt, and R. Ayyanar. "Probabilistic power flow studies for transmission systems with photovoltaic generation using cumulants." *IEEE Transactions on Power Systems* 27 (2012): 2251–2261.
6. F. J. Ruiz-Rodriguez, J. C. Hernandez, and F. Jurado. "Probabilistic load flow for radial distribution networks with photovoltaic generators." *IET Renewable Power Generation* (2012): 110–121.
7. S. Sannigrahi, S. R. Ghatak, and P. Acharjee. "Multi-scenario based bi-level coordinated planning of active distribution system under uncertain environment." *IEEE Transactions on Industry Applications* 56, no. 1 (2020): 850–863.
8. P. Kayal, and C. K. Chanda. "Optimal mix of solar and wind distributed generations considering performance improvement of electrical distribution network." *Renewable Energy* 75 (2015): 173–186.
9. D. Q. Hung, N. Mithulananthan, and R. C. Bansal. "Integration of PV and BES units in commercial distribution systems considering energy loss and voltage stability." *Applied Energy* 113 (2014): 1162–1170.
10. Y. Zheng, D. J. Hill, and Z. Y. Dong. "Multi-agent optimal allocation of energy storage systems in distribution systems." *IEEE Transactions on Sustainable Energy* 8 no. 4 (2017): 1715–1725.
11. M. M. Samy, S. Barakat, and H. S. Ramadan. "A flower pollination optimization algorithm for an off-grid PV-Fuel cell hybrid renewable system." *International Journal of Hydrogen Energy* 44 (2019): 2141–2152.
12. S. Singh, P. Chauhan, M. A. Aftab, I. Aftab, S. M. Hussain, and T. S. Ustun. "Cost optimization of a stand-alone hybrid energy system with fuel cell and PV." *Energies* 13 (2020): 1295.
13. S. Singh, P. Chauhan, and N. Singh. "Capacity optimization of grid connected solar/ fuel cell energy system using hybrid ABC-PSO algorithm." *International Journal of Hydrogen Energy* 45(February 23, 2020): 10070–10088.
14. J. Pascual, J. Barricarte, P. Sanchis, and L. Marroyo. "Energy management strategy for a renewable-based residential microgrid with generation and demand forecasting." *Applied Energy* 158 (2015): 12–25.
15. S. Jeyadevi, S. Baskar, C. K. Babulal, and M. W. Iruthayarajan. "Solving multiobjective optimal reactive power dispatch using modified NSGA-II." *International Journal of Electrical Power & Energy Systems* 33, no. 2 (2011): 219–228.
16. S. R. Ghatak, S. Sannigrahi, and P. Acharjee. "Multi-objective framework for optimal integration of solar energy source in three-phase unbalanced distribution network." *IEEE Transactions on Industry Applications* 56, no. 4 (2020): 3068–3078.
17. Andres Julian Aristizabal, Henry Giovanni Pinilla, and Carlos Andrés Forero. "Modeling of distributed power systems in 13 nodes IEEE electric grids." *Periodicals of Engineering and Natural Sciences* 4, no. 2 (2016).

18. M. Aien, M. G. Khajeh, M. Rashidinejad, and M. Fotuhi-Firuzabad. "Probabilistic power flow of correlated hybrid wind-photovoltaic power systems." *IET Renewable Power Generation* 8, no.6 (2014): 649–658.
19. S. Sannigrahi, S. Roy Ghatak, and P. Acharjee. "Strategically incorporation of RES and DSTATCOM for techno-economic-environmental benefits using search space reduction-based ICSA." *IET Generation, Transmission & Distribution* 13 no. 8 (2019): 1369–1381.
20. H. B. Tolabi, M. H. Ali, and M. Rizwan. "Simultaneous reconfiguration, optimal placement of DSTATCOM, and photovoltaic array in a distribution system based on fuzzy-ACO approach." *IEEE Transactions on SustainableEnergy* 6, no. 1. (2015):210–218.

6 Trends in Microgrid Control

Anup Kumar Nanda, Babita Panda, Chinmoy Kumar Panigrahi, Arjyadhara Pradhan, and Naeem Hannoon

CONTENTS

6.1 Introduction 119
6.2 Microgrids and the Indian Market 121
6.3 Problems in Islanded Mode of Operation 121
    6.3.1 Complicated Control 121
    6.3.2 Energy Storage 121
    6.3.3 Sustainable Advancement 122
    6.3.4 Fault Detection 122
    6.3.5 Voltage and Power Balance 122
    6.3.6 Economic Dispatch 122
6.4 Features of Microgrid Control Systems 123
6.5 Control Strategies for Microgrids 124
    6.5.1 Droop Control 125
        6.5.1.1 V/f Control 125
        6.5.1.2 PQ Control 125
    6.5.2 Master-Slave Strategy 126
6.6 Control Policies in a Microgrid 127
    6.6.1 Centralized Control Scheme 127
        6.6.1.1 Grid-Level Control 127
        6.6.1.2 Central Processing Control 127
        6.6.1.3 Field-Level Control 127
    6.6.2 Decentralized Control Scheme 128
6.7 Challenges and Tools in Microgrid Power Quality 130
6.8 Conclusion 132
References 132

6.1 INTRODUCTION

There still exist many areas which are not feasible to access from conventional main grids. In such areas where uninterrupted power supply becomes a challenge, the evolution of small autonomous grids stands out as the best complement. The recent

DOI: 10.1201/9781003121626-6

revolutionary changes in power systems allowing renewable energy sources (RESs) to be mingled into the power supply mix have made microgrids more prominent. Microgrids have proven highly beneficial in remote areas where major power challenges persist in terms of transmission, distribution and availability. Currently, the depletion of traditional conventional power sources has already reached extremes of risk. Future generations are likely to be highly affected if new energy sources don't run posthumously into the power supply system. A lot of research has been suggested in the literature encompassing control techniques in microgrids. Since renewable energy sources have a tendency to deliver intermittent power, control of microgrids is a crucial task. The major responsibilities of a microgrid controller comprise voltage-frequency regulation, active-reactive power exchange, hassle-free transition of modes between grid-connected and islanded, uninterrupted supply to critical loads and so on. Islanded operation of microgrids, which are DC, is advantageous over grid-tied systems in terms of voltage frequency synchronization issues. Microgrids are an integral part of a carbon-free economy and reliable power systems. Different renewable energy sources can be integrated easily and efficiently with the normal AC as well as DC loads. The motivation behind this chapter is to chart the relevant research done in the case of control of microgrids. A schematic diagram of a microgrid connected to loads is shown in Figure 6.1. Although different strategies have been deployed to promote microgrid and smart grid technology, many unanswered challenges still restrict the seamless growth of this technology. This chapter focuses on the existing challenges and trends in microgrid control and protection.

The rest of this chapter is organized as follows. Section 6.2 briefly summarizes the state of microgrids in Indian markets. In Section 6.3, various problems related to islanded operation of microgrids are discussed. Sections 6.4 and 6.5 focus on the

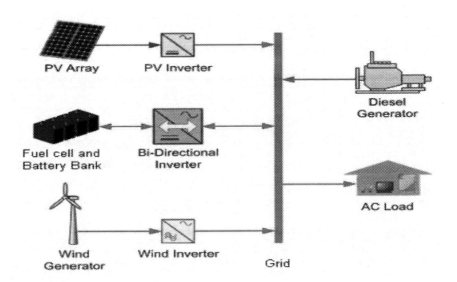

FIGURE 6.1 Microgrid connected to loads.

requisites and techniques of microgrid control systems. Section 6.6 discusses hierarchical control architecture for microgrids. In Section 6.7, some major persisting challenges in microgrid control are presented.

6.2 MICROGRIDS AND THE INDIAN MARKET

India has the largest network of integrated power in the world. Also, the ever-growing population raises a pertinent question about conventional power systems in terms of load management and uninterrupted power supply. The literature reveals an average loss as high as 26% associated with conventional systems. On this front, microgrids can be a feasible alternative pertaining to economic electrification over widespread regions. Keeping in view the per capita income and paying capability of Indian consumers, microgrids prove a comfortable option, ensuring economic and uninterrupted power.

6.3 PROBLEMS IN ISLANDED MODE OF OPERATION

In islanded state, the microgrid works separately without any connection with the power grid. So in order to provide persistent supply to loads, in this mode, control and protection risks are excessive. The demand may alter at times when the islanded mode of the microgrid is at risk. So proper load-sharing schedules should be accordingly implemented to avoid any cutoff in power supply, while grid-connected mode has a choice of bidirectional power flow between the powergrid and respective microgrid [1,2,3]. Therefore demand management can be done adequately with respect to the power grid in grid-connected operation. But when the power grid fails, major threats have also come up in order to provide consistent supply per demand as well as control of the integration of various renewable energy systems connected to the grid and from the protection point of view of the microgrid.

6.3.1 COMPLICATED CONTROL

When a large number of microgrids are connected as supply systems to the distribution grid, control becomes critical. Since microgrids are interfaced with a plethora of renewable energy sources and they are not connected to the host grid, control becomes more critical. So different microgrids need to be controlled by different DER units [4]. Integration of DERs in the system helps in reducing the physical and electrical distance between generation and loads. A local control approach based on the local measurements at the microgrid and a central control approach based on the cumulative and simultaneous function of several microgrids can be considered precisely to reach the power-demand balance [5]. Appropriate load dispatch schemes should be pre-decided, and loads should be supplied accordingly.

6.3.2 ENERGY STORAGE

Popularly, microgrids are modeled with solar PV energy or wind energy. But the major drawback of supply systems is the intermittent nature of power generation. Since the systems are dependent on the environment and weather conditions, there is no guarantee of a stable supply of power. During system disturbances, there is a major impact on the distributed energy storage systems interfaced with the microgrids. The response and health of these energy storage systems need to be studied carefully to stabilize the voltage-frequency and demand-generation. The literature suggests that a distributed energy storage system can cater to the consequences of the fluctuating supply of renewable energy sources [6, 7, 8, 9].

6.3.3 Sustainable Advancement

When the economics of microgrids is taken into account combining several cost-determining factors such as fuel expenses, maintenance costs, running capital, labor costs and so on, it can be supposed that energy generation from renewable sources is economic and more cost effective than traditional energy generation processes. The introduction of power electronic devices has added to the reduction of hardware and manpower as well, resulting in a reduction of overall cost of energy generation as compared to conventional power systems. If microgrids are planned and managed properly, then it can lead to substantial development and sustainable advancement of microgrid technology in the near future [10, 11, 12].

6.3.4 Fault Detection

As a variety of energy sources are interfaced with microgrids, the operation of different microgrids is different. In the case of any system faults, the traditional detection system fails to locate the fault exactly. In conventional power systems, relay technology is used for the purpose. But for microgrids, advanced and new detection techniques need to be effectively employed to mitigate the faults in the system without affecting the connected loads [13].

6.3.5 Voltage and Power Balance

Since different energy sources generate electricity at different levels of voltage, an eye must be kept on the frequency of the output voltage and power of the microgrids. Also, the microgrid must be able to handle any excess or shortfall of power without any reference to the host grid. Outputs of individual units must be traced and matched with the proper reference values to keep the deviations within safe permissible limits [14].

6.3.6 Economic Dispatch

A microgrid can be made to feed either purely AC or purely DC loads. Schematic diagrams of a typical DC microgrid and AC microgrid are shown in Figures 6.2 and 6.3, respectively. Scheduling the microgrid optimally to dispatch and manage connected loads and corresponding demands helps to reduce operation cost greatly. In islanded mode, reliability conditions need to be taken care of seriously and critically. Load-sharing schedules can be used in conjugation with optimal load management strategies to grow profit [15].

Trends in Microgrid Control

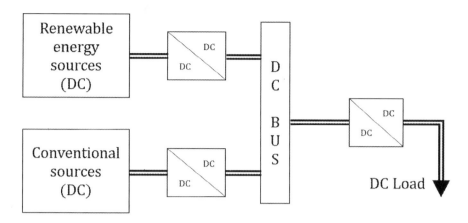

FIGURE 6.2 Schematic representation of DC microgrid.

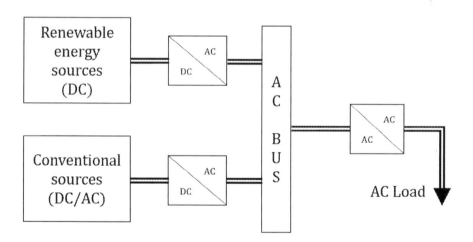

FIGURE 6.3 Schematic representation of AC microgrid.

6.4 FEATURES OF MICROGRID CONTROL SYSTEMS

For safe and steady operation of an islanded microgrid while facing many challenges, some of the desirable features of the control system are as follows.

- **Output Control**: The output voltage and currents of various distributed generations should be followed accurately to the nominal reference values. Prompt actions must be taken in the case of any mismatch found in the system [16].
- **Power Control**: A microgrid must adopt techniques to provide balance between the active and reactive power in the system. Individual control of

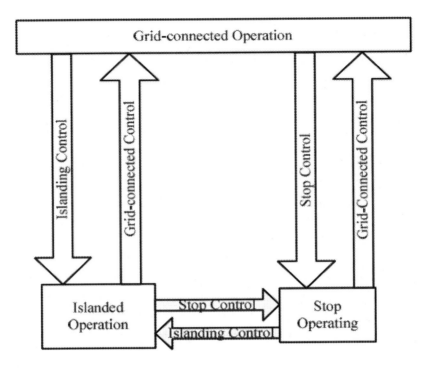

FIGURE 6.4 Operating modes of microgrids.

power is required at times of extra generation or power deficit at different DERs without altering the equilibrium of the system [17].
- **Lossless Distribution Networks**: Distribution of microgrid power according to the demand of loads is an essential requirement. However, in isolated areas where microgrids are the main supply of energy, cheap and lossless distribution is an important aspect of the control architecture.
- **Autonomous Operation**: The control technique must facilitate the microgrid working separately, whether in grid-connected or islanded mode. Autonomous action allows the microgrids to switch between various operational modes as well as a choice of convenient operating point [18,19].
- **Economic Dispatch**: Since microgrids supply power to remote areas, it is a important requirement to maintain the loads efficiently. A prior load management plan including a prompt response to demand absolutely decreases the operating and running charges in the microgrid. Moreover, economic dispatch aids the system in the long run by largely decreasing maintenance prices.

6.5 CONTROL STRATEGIES FOR MICROGRIDS

Renewable energy sources have evolved as a faithful alternative to fast-depleting conventional fuels. Due to the abundance of availability and their eco-friendliness, the integration of RES with microgrids has made strides in the era of ever-growing electricity demands. The traditional power flow techniques involve a unidirectional flow of power. But this mechanism is not suitable for the power flow associated with microgrids. In standalone mode, the slack bus is not taken into account [20,21,22]. Since the power obtained from RES-enabled microgrids is intermittent in nature, the grid frequency is taken to be a variable.

6.5.1 Droop Control

Droop control technique tracks the droop characteristics of the generation unit. At times of inadequate generation, droop control effectively adjusts the shading of the microgrid as required. The conventional P-ω and Q-V droop techniques have several disadvantages, as follows.

- As the load dynamics are not taken into account, this may lead to sudden power cutoff in the case of demand increase or decrease.
- The uses of power electronic devices introduce harmonics to the output of the microgrids so traditional schemes fail to manage non-linear loads.
- The basic regulation of any control scheme is to establish an overall system frequency to which the connected loads report. But traditional methods fail to establish the system frequency with dynamic loading conditions.

However, the new droop control manifesto is preferable over traditional schemes in terms of enhanced flexibility and prompt response to any disturbance. The lion's share of the control scheme depends on local measurements at the microgrids. This wipes out the requirement for a complex communication architecture, and there is hardly any interdependency among the local controllers. For islanded mode of microgrids, in the literature, two types of droop control schemes have been suggested [23,24,25].

6.5.1.1 V/f Control

The voltage generated by microgrids is stochastic in nature. Also, due to the presence of power electronic architecture in the system, harmonics appear at every instance. The grid frequency is taken as a variable in this method. A particular reference is set by the control mechanism for output voltage of the distributed generations and frequency of the grid. Due to mismatch in voltage level and frequency, oscillations occur in the system. The magnitude of oscillations directly affects the capacitance of the voltage control oscillator. Fluctuations in capacitance values of the voltage-controlled oscillator result in variable frequency. Droop characteristics of the load determine the amplitude of fluctuations. Voltage control ensures the amplitude of the voltage at a particular level by adjusting the reactive power output of the distributed generations. The frequency control adjusts the active power output of the distributed generations [26,27,28].

6.5.1.2 PQ Control

The microgrid central controller sets a nominal value as reference for the active and reactive power generated by the microsource. This method ensures a constant active and reactive power output of the microsource. When the microgrid operates in

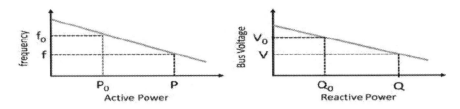

FIGURE 6.5 Droop characteristics of distributed generation.

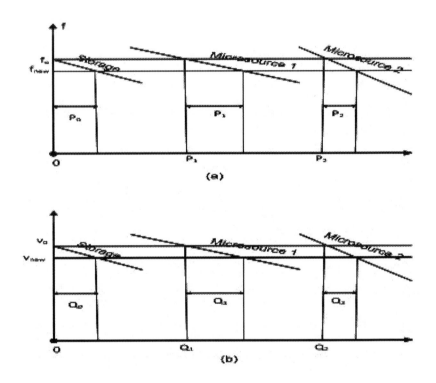

FIGURE 6.6 V/f control: (a) voltage control, (b) frequency control.

islanded mode, this control technique is deployed for distributed power supply, where a constant power level is maintained at the time of dispatch. In standalone operations, the distributed power system is taken as the main control unit. The PQ control technique is applied, followed by V/f control [13,29,32].

6.5.2 Master-Slave Strategy

In islanded operation of microgrids, the distributed power is considered the master distributed generation. The strategy for output voltage and frequency controls the other distributed generations interfaced with the master. Usually the master distributed generation carries the main power. The controlled coordinated power flows through all other distributed generations. This strategy employs voltage-frequency control followed by active-reactive power control for other distributed generations [36, 37].

6.6 CONTROL POLICIES IN A MICROGRID

Generally, two approaches are tried with microgrids to achieve faithful and reliable control: centralized and decentralized. A centralized control scheme operates on global data of different distributed generations accumulated centrally at a control processing unit. This scheme requires an extensive communication network to collect the instantaneous data and employ the optimal control commands. The communication network replicates the human sensory system. The central cluster behaves as the brain, which gathers all data from different controllers simultaneously and consistently through the nerves and sensory neurons. Then it is on the part of the brain of the network to process the information and appropriately decide the control variables and their corresponding control signals for each isolated controller [18, 22, 25, 26]. Contrary to this, the decentralized control method depends upon local data collected at each distributed generation to control the unit. Each component operates autonomously, and close coordination between the individual components is necessary. A decentralized system is unaware of the changing of control variables.

6.6.1 CENTRALIZED CONTROL SCHEME

The literature suggests a three-tier policy for the centralized control scheme: grid-level, central processing-level and field-level control.

6.6.1.1 Grid-Level Control

This level decides the energy dispatch policy for the microgrid according to market economic stands. The energy generation schedule and distribution schedule are stacked in this level. The optimal balance of active reactive power is preset at this stage before opening up to distribution.

6.6.1.2 Central Processing Control

This stage is the hub for all the local data received from distributed generations. A microgrid central controller (MGCC) matches the generated voltage level and frequency with the preset reference values. Accordingly, an error signal is advanced to the generation modules to satisfy the required levels of active and reactive power in the system. Decisions in emergency situations regarding load shedding, blackouts or black restoration are identified in this level of control [28, 30,38,39].

6.6.1.3 Field-Level Control

Each distributed generation unit is provided with a local controller which independently operates the control variables based upon local measurements and calculations. This level is more reliable, as it does not involve a communication network. Droop control is adopted to manipulate the control variables [31, 33].

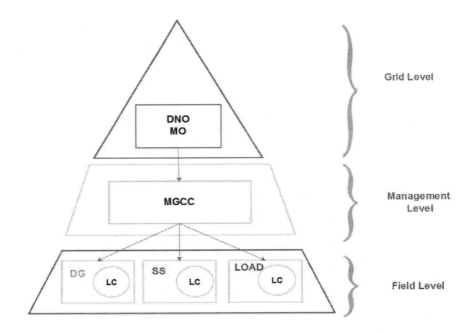

FIGURE 6.7 Centralized control policy for microgrids.

6.6.2 Decentralized Control Scheme

This scheme doesn't involve a central processing unit. Rather each module has its own local controller. Different control variables are addressed for different units. No common control command flows over the system. This method is more popular in islanded operation of microgrids, as it does not require a communication network. The distribution network is classified as small individual units. Each unit operates autonomously and adheres to optimal operating conditions. This control requires fewer observed variables and acts efficiently by observing fewer parameters to issue control commands to the corresponding unit. All the decisions made by the controller are communicated, irrespective of systemwide changes.

Power systems extending over larger areas are hard to handle with decentralized control, as the distance between the components increases the number of messages communicated between the controller and system, thereby requiring a dedicated communication network. Also, the coupling between active and reactive power in the system leads to problems. So a midway policy is adopted to address such situations. The midway policy involves three stages of control: primary, secondary and tertiary. These levels are distinguished from each other in terms of speed of response, communication requirements and reference set values [34, 35]. A general control hierarchy is shown in the following figures.

Trends in Microgrid Control

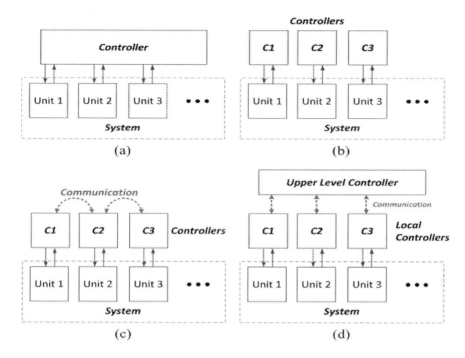

FIGURE 6.8 Control strategies: (a) centralized, (b) decentralized, (c) distributed, (d) hierarchical.

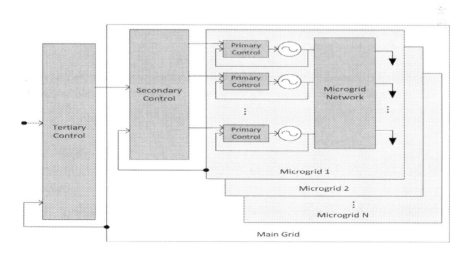

FIGURE 6.9 Hierarchical control in microgrids.

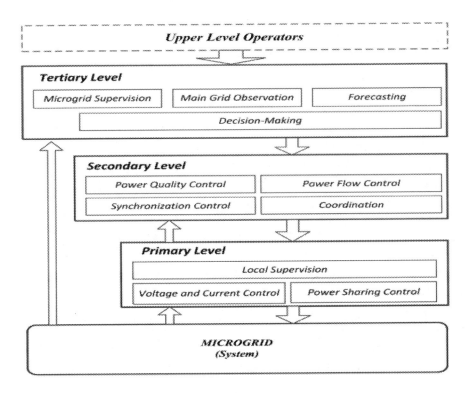

FIGURE 6.10 Multilevel control of microgrids.

The main functions of the multilevel control techniques are as follows.

- Primary level targets local measurement of different control variables. Local voltage and current control is attained in this level [40, 43].
- Secondary control gets information from primary control. It focuses on power quality control, coordination among different sub-modules, power flow control and so on. This level refers to the microgrid energy management system (MEMS). This level is responsible for the optimal dispatch and economic working of the microgrid [41, 44, 46].
- Tertiary control is the top control stage, which directs the microgrid as a whole. This control level establishes long-term set points known as the optimum operating range for the whole microgrid unit. It takes care of the total active and reactive power balance [47, 48].

6.7 CHALLENGES AND TOOLS IN MICROGRID POWER QUALITY

Maintaining the quality of power delivered to the loads is as important as generation. The IEEE has set certain standards for the quality of power associated with microgrids. Moreover, the imbalance in load demand creates challenges for

Trends in Microgrid Control

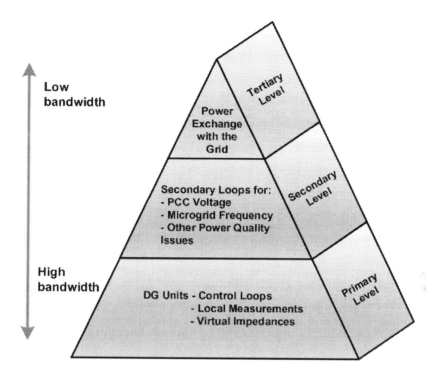

FIGURE 6.11 Control architecture of microgrids.

microgrids, especially in islanded mode of operation [42, 45]. Some of the major challenges are given in the following:

- Integration of renewables.
- Integration of electric vehicles.
- Use of power electronic components with a limit on the harmonics injected into the system.

Different tools to satisfy standardized power quality are as follows:

- Modern monitoring devices
- Smart metering facilities
- Smart appliances
- Storage devices
- Advanced information and communication technology
- Computational intelligence
- Advanced control techniques
- Efficient demand management
- Active demand response
- Economic load dispatch schedules

6.8 CONCLUSION

In recent years, microgrids have evolved as a hot area of research. In comparison to conventional wide-area power systems, the control variables in microgrids are available in plenty. In parts of the United States and Europe, research is at a peak to make microgrids more stable, economic and environmentally friendly. For remote electrification, microgrids are more reliable and steady because they have fewer observed parameters for each control scheme. The development of microgrids has geared up because of projects like China's 863 and 973 development projects, and they stand as an alternative to traditional power systems. The advancement of microgrids has driven the demand for implementation of various renewable energy resources like wind, PVs and battery and fuel cells into mainstream power scenarios. For conventional microgrids, droop control is determined to be the universally used control technique, while for islanded mode, voltage frequency control is widely used to control microgrids. A hierarchical control scheme is noticeable in the literature, playing an important role in control of islanded microgrids by breaking down the whole system into small modules. The most general control strategy consists of two stages, centralized and decentralized control. The PQ control method focuses on power coupling problems in islanded microgrids. Other control strategies include master-slave strategy and peer-to-peer control strategy for islanded microgrids; V/f and PQ control methods are discussed where study is limited. This chapter presents a cumulative overview of various control trends for microgrids. Future research in the field can lead the process of transforming conventional power systems into smart power systems. Ongoing research has reached the halfway pointin the realization of smart grids as an alternative to traditional power systems.

REFERENCES

1. F. Mumtaz, M. H. Syed and M. Al Hosani, et al., "A simple and accurate approach to solve the power flow for balanced islanded microgrids," in IEEE 15th International Conference on Environment and Electrical Engineering (EEEIC), pp. 1852–1856, 2015.
2. J. Lopes, C. Moreira and A. Madureira, "Defining control strategies for microgrids islanded operation," *IEEE Transactions on Power Systems*, vol. 21, pp. 916–924, May 2016.
3. IEEE guide for design, operation, and integration of distributed resource island systems with electric power systems, IEEE Std. 1547.42011, pp. 1–54, 2011.
4. T. Dragičević, X. Lu, J. C. Vasquez, et al., "DC microgrids—part I: A review of control strategies and stabilization techniques," *IEEE Transactions on Power Electronics*, vol. 31, no. 7, pp. 4876–4891, July 2016.
5. T. Dragičević, X. Lu, J. C. Vasquez, et al., "DC microgrids—part II: A review of power architectures, applications, and standardization issues," *IEEE Transactions on Power Electronics*, vol. 31, no. 5, pp. 3528–3549, May 2016.
6. R. Zamora and A. K. Srivastava, "Controls for microgrids with storage: Review, challenges, and research needs," *Renewable and Sustainable Energy Reviews*, vol. 14, no. 7, pp. 2009–2018, September 2010.
7. R. Zamora and A. K. Srivastava, "Controls for microgrids with storage: Review, challenges, and research needs," *Renewable and Sustainable Energy Reviews*, vol. 14, no. 7, pp. 2009–2018, September 2010.

8. T. Dragičević, X. Lu, J. C. Vasquez, et al., "Tertiary and secondary control levels for efficiency optimization and system damping in droop controlled DC–DC converters," *IEEE Transactions on Smart Grid*, vol. 6, no. 6, pp. 2615–2626, November 2015.
9. A. Bidram and A. Davoudi, "Hierarchical structure of micro grids control system," *IEEE Transactions on Smart Grid*, vol. 3, no. 4, pp. 1963–1976, December 2012.
10. W. Chengshan and L. Peng, "Development and challenges of distributed generation, the micro-grid and smart distribution system[J]," *Automation of Electric Power Systems*, vol. 34, no. 2, pp. 10–14, 2010.
11. W. Guo-dong, "Overview on smart micro-grid research [J]," *China Electric Power (Technology Edition)*, vol. 2, p. 021, 2012.
12. T. Caldognetto, S. Buso, P. Tenti, et al., "Power-based control of low-voltage microgrids," *IEEE Journal of Emerging and Selected Topics in Power Electronics*, vol. 3, no. 4, pp. 1056–1066, 2015.
13. D. I. Brandao, T. Caldognetto, F. P. Marafao, et al., "Centralized control of distributed single-phase inverters arbitrarily connected to three-phase four-wire microgrids," *IEEE Transactions on Smart Grid*, vol. 8, no. 1, pp. 437–446, 2017.
14. A. Micallef, M. Apap, C. Spiteri-Staines, et al., "Single-phase microgrid with seamless transition capabilities between modes of operation," *IEEE Transactions on Smart Grid*, vol. 6, no. 6, pp. 2736–2745, 2015.
15. A. Milczarek, M. Malinowski and J. M. Guerrero, "Reactive power management in islanded microgrid—proportional power sharing in hierarchical droop control," *IEEE Transactions on Smart Grid*, vol. 6, no. 4, pp. 1631–1638, 2015.
16. X. Chen, Y. Hou, S. C. Tan, et al., "Mitigating voltage and frequency fluctuation in microgrids using electric springs," *IEEE Transactions on Smart Grid*, vol. 6, no. 2, pp. 508–515, 2015.
17. Z. Akhtar, B. Chaudhuri, S. Y. Ron Hui, et al., "Primary frequency control contribution from smart loads using reactive compensation," *IEEE Transactions on Smart Grid*, vol. 6, no. 5, pp. 2356–2365, 2015.
18. T. L. Vandoorn, J. C. Vasquez, J. De Kooning, et al., "Microgrids: Hierarchical control and an overview of the control and reserve management strategies," *IEEE Industrial Electronics Magazine*, vol. 7, no. 4, pp. 42–55, 2013.
19. H. Han, X. Hou, J. Yang, et al., "Review of power sharing control strategies for islanding operation of AC microgrids," *IEEE Transactions on Smart Grid*, vol. 7, no. 1, pp. 200–215, 2016.
20. L. Meng, E. R. Sanseverino, A. Luna, et al., "Microgrid supervisory controllers and energy management systems: a literature review," *Renewable and Sustainable Energy Reviews*, vol. 60, pp. 1263–1273, 2016.
21. Y. Han, H. Li, P. Shen, et al., "Review of active and reactive power sharing strategies in hierarchical controlled microgrids," *IEEE Transactions on Power Electronics*, vol. 32, no. 3, pp. 2427–2451, 2017.
22. M. Yazdanian and A. Mehrizi-Sani, "Distributed control techniques in microgrids," *IEEE Transactions on Smart Grid*, vol. 5, no. 6, pp. 2901–2909, 2014.
23. A. Milczarek, M. Malinowski and J. M. Guerrero, "Reactive power management in Islanded microgrid—proportional power sharing in hierarchical droop control," *IEEE Transactions on Smart Grid*, vol. 6, no. 4, pp. 1631–1638, 2015.
24. H. Mahmood, D. Michaelson and J. Jiang, "Reactive power sharing in islanded microgrids using adaptive voltage droop control," *IEEE Transactions on Smart Grid*, vol. 6, no. 6, pp. 3052–3060, 2015.
25. A. Bidram and A. Davoudi, "Hierarchical structure of microgrids control system," *IEEE Transactions on Smart Grid*, vol. 3, no. 4, pp. 1963–1976, December 2012.
26. Z. Lecai, "Technology development research of distributed energy and the micro-grid [J]," *Journal of Shanghai Electric Technology*, vol. 1, p. 013, 2013.

27. J. M. Guerrero, P. C. Loh and T. L. Lee, et al., "Advanced control architectures for intelligent microgrids—part II: Power quality, energy storage, and AC/DC microgrids," *IEEE Transactions on Industrial Electronics*, vol. 60, no. 4, pp. 1263–1270, 2013.
28. Thillainathan Logenthiran, Ramasamy Thaiyal Naayagi, Wai Lok Woo, Van-Tung Phan and Khalid Abidi, "Intelligent control system for microgrids using multiagent system," *IEEE Journal of Emerging and Selected Topics in Power Electronics*, vol. 3, no. 4, 2015.
29. A. Mehrizi-Sani and M. R. Iravani, "Potential-function based control of a microgrid in islanded and grid-connected modes," *IEEE Transactions on Power Systems*, vol. PP, no. 99, pp. 1–8, 2010.
30. J. Y. Kim, et al., "Cooperative control strategy of energy storage system and microsources for stabilizing the microgrid during islanded operation," *IEEE Transactions on Power Electronics*, vol. 25, no. 12, pp. 3037–3048, December 2010.
31. Z. Fan, et al., "Smart grid communications: Overview of research challenges, solutions, and standardization activities," *IEEE Communications Surveys and Tutorials*, vol. 15, no. 1, pp. 21–38, 1sr Quart. 2012.
32. S. M. Amin, "For the good of the grid," *IEEE Power & Energy Magazine*, vol. 6, no. 6, pp. 45–89, 2008.
33. N. Hatziargyriou, H. Asano, R. Iravani and C. Marnay, "Microgrids," *IEEE Power & Energy Magazine*, vol. 5, no. 4, pp. 78–94, 2007.
34. F. Katiraei, R. Iravani, N. Hatziargyriou and A. Dimeas, "Microgrids management," *IEEE Power & Energy Magazine*, vol. 6, no. 3, pp. 54–65, 2008.
35. A. Mehrizi-Sani and M. R. Iravani, "Potential-function based control of a microgrid in islanded and grid-connected modes," *IEEE Transactions on Power Systems*, vol. PP, no. 99, pp. 1–8, 2010.
36. W. Guo-dong, "Overview on smart micro-grid research," *China Electric Power (Technology Edition)*, vol. 2, p. 021, 2012.
37. P. Kundur, *Power System Stability and Control*, McGraw-Hill, New York, 1994.
38. R. Zamora and A. K. Srivastava, "Controls for microgrids with storage: Review, challenges, and research needs," *Renewable and Sustainable Energy Reviews*, vol. 14, no. 7, pp. 2009–2018, September 2010.
39. S. Bahramirad and W. Reder, "Islanding applications of energy storage system," in Power and Energy Society General Meeting, 2012 IEEE, pp. 1–5, 2012.
40. T. Xisheng and Q. Zhiping, "Energy storage control in renewable energy based microgrid," in Power and Energy Society General Meeting, IEEE, pp. 1–6, 2012.
41. T. L. Vandoorn, B. Renders, L. Degroote, et. al., "Active load control in islanded microgrids based on the grid voltage," *IEEE Transactions on Smart Grid*, vol. 2, pp. 139–151, 2011.
42. T. L. Vandoorn, B. Meersman, L. Degroote, et. al., "A control strategy for islanded microgrids with DC-link voltage control," *IEEE Transactions on Power Delivery*, vol. 26, pp. 703–713, 2011.
43. M. Pipattanasomporn, H. Feroze and S. Rahman, "Securing critical loads in a PV-based microgrid with a multi-agent system," *Renewable Energy*, vol. 39, pp. 166–174, 3// 2012.
44. T. Hiyama, T. Nagata and T. Funabashi, "Multi-agent based operation and control of isolated power system with dispersed power sources including new energy storage device," in Transmission and Distribution Conference and Exhibition 2002: Asia Pacific, IEEE/PES, 2002.
45. Y. J. Reddy, S. Dash, A. Ramsesh, et. al., "Monitoring and control of real time simulated microgrid with renewable energy sources," in Power India Conference, 2012 IEEE Fifth, pp. 1–6, 2012.

46. S. Bahramirad and W. Reder, "Islanding applications of energy storage system," in Power and Energy Society General Meeting, 2012 IEEE, pp. 1–5, 2012.
47. P. Basak, A. K. Saha, S. Chowdhury, et. al., "Microgrid: Control techniques and modeling," in Universities Power Engineering Conference (UPEC), 2009 Proceedings of the 44th International, pp. 1–5, 2009.
48. T. L. Vandoorn, B. Meersman, L. Degroote, et. al., "A control strategy for islanded microgrids with DC-link voltage control," *IEEE Transactions on Power Delivery*, vol. 26, pp. 703–713, 2011.

7 Utility Tariff Variation and Demand Response Events
A Case Study of Microgrid Design and Analysis

Aashish Kumar Bohre and Parimal Acharjee

CONTENTS

7.1	Introduction	138
7.2	Solar Power Resource Assessment for Case Study	140
7.3	Modeling of Microgrid System	140
	7.3.1 Solar-PV	141
	7.3.2 Battery Storage System	142
	7.3.3 System Converter	142
	7.3.4 Diesel Generator	143
	7.3.5 Load	143
7.4	Description of Problem and Methodology	144
	7.4.1 Reliability of System	147
	7.4.2 Resilience of System	148
	7.4.3 Demand Response	148
	7.4.4 Total Net Present Cost	148
	7.4.5 Levelized Cost of Energy	149
	7.4.6 Simple Payback	149
	7.4.7 Pollutant Emissions and Particulate Matter	149
7.5	Results and Discussion	150
	7.5.1 Base Case System (Current Existing System of NITD): Diesel Generator Connected to DVC-NITD-Grid	150
	7.5.2 Proposed Microgrid System: Solar PV-Battery Storage-Diesel Generator Connected to Main Utility Grid	159
7.6	Conclusions	177
7.7	Acknowledgments	178
References		178

DOI: 10.1201/9781003121626-7

7.1 INTRODUCTION

Nowadays, renewable energy sources are more auspicious in power generation to develop a more efficient and reliable microgrid system based on the renewable energy system due to the exhaustible nature of fossil fuels and global warming. One of the most abundant renewable sources available in nature is the solar PV. Therefore, the combination of solar PVs with energy storage systems, including generators as power backup-based microgrids connected to the main grid, has great potential to provide better-quality and more reliable power to customers than a conventional system based on a single utility. At present, the extraction of renewable energy sources for power generation systems has gained more attention worldwide. Microgrids may be standalone, an off-grid type or a grid-connected type, and they are self-sustainable energy systems and can maintain system reliability and resiliency. The standalone, off-grid or isolated microgrid requires sufficient storage and a power backup system to maintain reliability. The grid-connected microgrid has the capability to supply power to local loads as well as to the grid by employing net metering or net purchased power from the grid approach. Paterakis et al. [1] presented an overview of a demand response (DR) program while utilizing renewable energy sources connected to the utility and also gave information about incentive and price-based DR event programs to accomplish various economic, technical and operational benefits. Chai et al. [2] analyzed a demand response model to maximize the benefits of power retailers based on incentive rates or rebates when the behavior of customer load demand/consumptions changes. The more cost-efficient renewable hybrid microgrid system is achieved by employing solar and wind energy sources, and optimal incentive charges are attained based on the DR model. He et al. [3] implemented a renewable energy-based microgrid system for sustainable operation of a large-scale residential community in China for grid-connected and standalone mode operation of a microgrid using HOMER to obtain economic performance of various system components and cost-effective design among different configurations. Chrysopoulos et al. [4] applied cost-effective demand response for small-scale consumers based on prediction of the customers' individual activities. The planned prototype was also applied to forecast the load demand through which pricing policies can be modified based on customer comfort and partialities. An emergency demand response based on the reliability optimization method by considering generation failure was executed by Aghaei et al. [5], which positively supported system reliability and other technical and economic facilities for emergency operation. Tang et al. [6] employed a load control scheme for demand response used to manage power within the smart grid by performing data center load management with DR through a data center and smart grid application. Babar et al. [7] performed an incentive bid-based demand reduction algorithm by direct load control in combined demand response plans which minimize the congestion of the network to decrease the available generation and load demand mismatch. Masrur et al. [8] proposed a techno-economic-environmental analysis of an isolated microgrid system in Bangladesh Island, which concluded that the proposed hybrid microgrid system based on solar PV and wind was efficient to reduce the high per-unit cost of energy and increase the economic and technical feasibility of microgrids. Safamehr et al. [9] realized microgrid performances based on cost effectiveness and sustainable energy management with an intelligent demand response approach using an artificial bee colony and quasi-static algorithm to minimize the total cost of energy of the microgrid system. Different configuration planning for microgrids and techno-economic analysis consisting of a PV and wind system for rural-area electrification of parts of Indian states were presented using a different strategy by Suresh et al. [10] and Mazzola et al. [11]. The optimal design and analysis of a nano-grid system based on a solar PV-wind-battery considering laboratory experimentation was given by Tudu et al. [12], which provides an efficient microgrid planning path for practical systems. Zhang et al. [13] scrutinized the optimal design of CHP-based microgrids, including renewable distributed energy resources like solar, using a multiobjective optimization approach to decrease the cost and environmental impacts.

A feasibility analysis of a standalone hybrid microgrid comprising PV, wind and other system components to find the optimal configuration of the microgrid and operation with respect to total cost and the number of energy generation sources was explored in Kumar et al. [14] and García et al. [15]. Thomas et al. [16] utilized different renewable sources with penetration impact for a small isolated hybrid microgrid system design by considering the different scenarios of a case study, and techno-economic analysis was performed for the case study. Sawle et al. [17] evaluated the modeling and analysis of a hybrid off-grid system based on renewable sources like solar and wind to investigate various social, technical and economic criteria such as loss of power supply probability (LPSP), cost economy, particulate matter (PM), emissions and human development using a particle swarm optimization technique for an off-grid system. Guo et al. [18] planned a standalone microgrid system for remote areas considering new multiple objectives for a distribution system using a pareto optimal solution-based approach to obtain different benefits from the planned system. An assessment of numerous potential benefits and motivations offered by hybrid renewable energy sources was given by Elavarasan et al. [19], which argued for more use of renewable sources globally. Recent development in the socio-economic area, motivations, issues, system configuration and planning methodologies using different objectives to evaluate the performance and operation of a hybrid renewable energy system in off-grid and grid-connected modes were analyzed by Babatunde et al. [20] using different optimization techniques and methodologies and also including the HOMER [21] simulation software platform. The planning and design of low-cost renewable energy solutions for various systems and microgrids based on the availability of renewable energy resources, reliability and sensitivity analysis to decide possible configurations of a system and its analysis were presented by Go et al. [22]. A multi-objective approach for optimal design and analysis of hybrid renewable energy systems with different system components like fuel cell-based micro-CHPs, electrolyzer PVs and wind systems using sunflower optimization was executed by Fan et al. [23]. Bohre et al. [24] investigated a multi-objective approach with system performance indices for the optimal planning of distribution systems with different types of distributed generations, including renewables, using optimization techniques. An experimental study to manage energy within standalone and grid-connected solar-wind based hybrid systems together with battery storage was implemented by Dali et al. [25] using a controlled interconnection concept of the presented grid. Saib et al. [26] considered a grid-connected hybrid renewable system comprising PV, wind and battery storage simulated in the MATLAB environment to access system performance information such as voltage profile, power injected to the grid and so on. The optimum planning and analysis of a grid-connected hybrid solar-wind-battery-based generation system to meet the optimal generation-demand basis of generic methodology was examined by Khalid et al. [27].

The previous summary and discussion of various microgrid case studies and hybrid renewable energy system design and optimization in different operating modes such as off-grid/standalone/isolated and grid-connected systems motivated using the benefits of a solar-based renewable microgrid system for the proposed case study. Many design and analysis tools and optimization algorithms are available, but the HOMER software tool and MATLAB are widely used and globally accepted by researchers. Hence, both HOMER-grid and MATLAB are utilized for the design and analysis of the proposed case study of National Institute of Technology (NIT), Durgapur. The detailed technical, economic and social benefits of the proposed grid-connected microgrid are presented using the HOMER-grid optimizer, including demand response and utility tariff options. The comparison is given with respect to the existing base case system.

7.2 SOLAR POWER RESOURCE ASSESSMENT FOR CASE STUDY

The availability of renewable solar resources for the proposed case study location, NIT Durgapur (NITD)was assessed by NASA's prediction of the Worldwide Energy Resource (Power) database with HOMER-grid. The details of the proposed case study location, as shown in Figure 7.1 for NITD, are as follows [21]:

- Location: Academic Building, NIT, A-Zone, Durgapur, West Bengal 713209, India
- Latitude: 23°33.3' N
- Longitude: 87°17.4' E
- Time zone: Asia/Kolkata

The topographical location and availability of annual solar data for the case study NITD are shown in Figures 7.1 and 7.2, respectively. The annual average solar global horizontal irradiance is 4.824 kW/m^2/day, which is given in Table 7.1. The available solar resources for the NITD case study are updated in HOMER-grid with the help of the MATLAB file interface. The final solar output power is calculated using HOMER-grid.

7.3 MODELING OF MICROGRID SYSTEM

The microgrid system model includes the PV array, battery storage, system converter and other essential components to maintain the reliability and resilience of the proposed grid-connected microgrid system, which is discussed in this section.

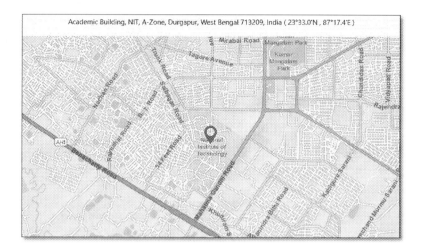

FIGURE 7.1 NITD topographical location.

TABLE 7.1
Monthly Average Solar Clearance Index Global Horizontal Irradiance, Temperature and Wind Speed

Month	Clearance Index	Radiation (kW/m²/day)	Temperature (°C)
Jan	0.609	4.230	16.960
Feb	0.622	5.010	21.130
Mar	0.614	5.770	26.880
Apr	0.595	6.230	31.660
May	0.555	6.120	32.830
Jun	0.446	4.980	31.010
Jul	0.394	4.350	28.750
Aug	0.400	4.230	28.120
Sep	0.429	4.140	27.200
Oct	0.530	4.430	25.040
Nov	0.601	4.290	21.070
Dec	0.624	4.110	17.300
Annual Average	0.535	4.824	25.663

7.3.1 Solar-PV

The solar-PV power output depends on the temperature and irradiation of the study area, which is determined by the following equation [10–12].

$$P_{PV}^t = P_{PV}^0 \times d_f \times \left(\frac{I_r^t}{I_r^{STC}} \right) \times \left\{ 1 + \tau_c \left(T^t - T^{STC} \right) \right\} \tag{7.1}$$

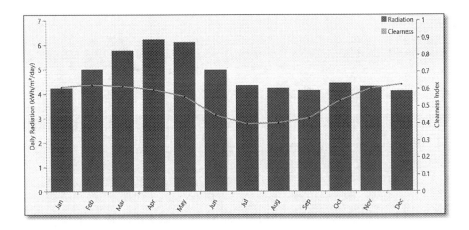

FIGURE 7.2 Availability of annual solar data for NITD.

where P_{PV}^t is the PV array power output at instant t; P_{PV}^0 is the PV array capacity for standard test conditions (STC); d_f is the derating factor of PV array in %; I_r^t is the solar radiation at time t in kW/m²; τ_c is the temperature coefficient; I_r^{STC} is the radiation at STC, that is, 1 kW/m²; T^t is temperature at time t in °C; and T^{STC} is the temperature at STC, that is, 25°C.

7.3.2 Battery Storage System

The excess energy generated by microgrid sources during light load conditions is stored in the battery storage system (BSS), and it dispatches suitable energy whenever required. The mathematical equation for battery state of charge during charging and discharging mode is given as [23, 25]:

$$B_{SOC}^t = B_{SOC}^{t-1} \pm \left(\frac{V_{bat}^t \times \eta_{bat}}{C_{bat}} \right) \qquad (7.2)$$

where the +ve and -ve signs are considered for charging and discharging mode operation of the battery energy storage system. Also, B_{SOC}^t and B_{SOC}^{t-1} are the battery state of charge for the t and $t-1$ instances, respectively; η_{bat} is the efficiency of the battery; V_{bat}^t is the discharged or charged power by the battery at instance t; and C_{bat} is the battery capacity in kWh. The BES can be charged up to B_{SOC}^{max} and discharged up to B_{SOC}^{min}, where the minimum and maximum SOC limits are specified as:

$$B_{SOC}^{min} \leq B_{SOC} \geq B_{SOC}^{max}$$

7.3.3 System Converter

The electronic system converter is required for maintaining the energy flow among the different AC and DC components. The electric energy is transferred from one

Utility Tariff Variation and Demand Response Events 143

element to the other by different operating modes of the system converter, that is, converters (AC to DC)/inverters (AC to DC) per the needed load frequency. The efficiency of the converter can be evaluated as [17]:

$$\eta_{inv} = \frac{P}{P + P_0 + kP^2} \tag{7.3}$$

where P, P_0 and k are correspondingly obtained as:

$$P_0 = 1 - 99\left(\frac{10}{\eta_{10}} - \frac{1}{\eta_{100}} - 9\right)^2, k = \frac{1}{\eta_{100}} - P_0 - 1, \text{ and } P = P_{out}/P_n \tag{7.4}$$

where η_{10} and η_{100} are the inverter's efficiency for 10% and 100% of rated energy, respectively.

7.3.4 DIESEL GENERATOR

The diesel generator set is used for alternate backup supply power in the system. The diesel generator fuel consumption is dependent on the size of generator and connected load to which the generator supplies the power. The diesel generator fuel consumption can be calculated as [23]:

$$F_{cons} = k_0 W_0 + k_1 W_{dg} \tag{7.5}$$

where k_0 and k_1 are fuel consumption coefficients, W_0 and W_{dg} are diesel generator rated capacity and output power. Also, $k_0 = 0.081451$ l/kwh and $k_1 = 0.2461$ l/kwh. The efficiency of the diesel generator is defined by the ratio of the power output to the fuel consumption heating value [23].

$$\eta_{dg} = \frac{W_{dg}}{F_{cons} \times V_{lh}} \tag{7.6}$$

where V_{lh} is the low heat value of gas/oil and ranges from 10 to 11.6 kwh/lit.

7.3.5 LOAD

The commercial load is considered in the study on the basis of daily energy consumption in kWh/day of NITD. The estimation of the average load of the microgrid system can be distinguished as the area under the load curve for the specified duration. The per day and annual energy demand of NITD are given in Figures 7.3 and 7.4, respectively. The peak load demand is 1,488.52 kW. The average daily, monthly and annual electric energy consumption/demand/load are 13,830.6 kWh/day, 420.7 MWh/month and 5,048.2 MWh, respectively. The total load of the NITD case study is divided into critical or primary load and non-critical load. The critical load is the primary load which requires a power supply for primary official work and a power supply in emergencies for laboratory, health center and other main priority work, which is supplied even in outages from a backup supply. The load model of the case

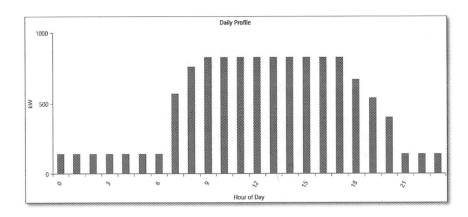

FIGURE 7.3 Load profile for a day at NITD.

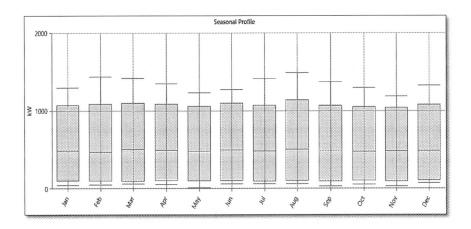

FIGURE 7.4 Seasonal load profile of NITD.

study is implemented in HOMER-grid with the help of an imported MATLAB file interface. The parameters considered for proposed grid-connected microgrid case study are given in Table 7.2.

7.4 DESCRIPTION OF PROBLEM AND METHODOLOGY

The base case of the presented work is the current existing system of the NITD, WB, India. The current system at NITD, including the AC load, diesel generator (DG) and main grid, is considered the base case system. The NITD system is connected to the main grid, which is connected to the Damodar Valley Corporation (DVC) power plant following the DVC tariff (tariff-1) order for energy consumption [28]. The microgrid comprising solar PV, battery storage and diesel generator is proposed for the NITD

TABLE 7.2
Different Components of Proposed Microgrid for Case Study

Solar-PV (Generic flat PV)	Diesel Generator	System Converter
Cost:	**Cost:**	**Cost:**
Capital: ₹ 3000/kW	Initial: ₹ 52,500/kW	Capital: ₹ 300/kW
O&M: ₹ 18/yr	Replacement: ₹ 35,000/kW	O&M: ₹ 10/yr
Replacement: ₹ 3000/kW	O&M: ₹ 0.7 /h	Replacement: ₹ 300/kW
Lifetime: 25 yr	Sizes Considered: 1500 kW	Efficiency: 90%
Sizes Considered: 300, 400 and 500kW	Maximum Load Ratio: 25%	Lifetime: 15 yr
Derating Factor: 80%	Lifetime: 15,000 hrs	**Rates:**
Battery Storage (Generic LA)	Diesel Fuel Price: ₹ 66.0/L	Discount: 8%
Cost:	Lower Heating Value: 43.2 MJ/kg	Inflation: 2%
Capital: ₹ 550/kWh	Density: 820.00 kg/m^3	Interest: 6%
O&M: ₹ 10/yr	Sulfur Content: 0.4%	Project Lifetime: 25 yr
Replacement: ₹ 550/kWh	Carbon Content: 88.0%	
Lifetime Throughput: 3000 (kWh)		
Nominal Voltage: 6 V		
SOC Min.: 20%		

system in this study, which offers a lower per-unit price of energy and net present cost. The pollution in Durgapur city is also increasing day by day due to many reasons such as the presence of large industries, use of conventional and fossil fuel-based power plants, internal combustion (IC) engine vehicles and many more. Thus, global warming, human health and other social impacts are adversely affected. The current need is to develop a green campus globally; therefore, a renewable source, solar PV, is proposed with the existing NITD grid-connected power system. The proposed microgrid integration will improve the system's social, economical and technical stability at a lower per-unit energy price and with flexible and reliable overall system operation. The energy generated by the proposed microgrid system, including PV, is utilized to meet the load demand with grid connection mode. The total load demand is served by the combined economic operation of the grid-connected microgrid including, PV panels, storage systems and DG unit. When the supply by the combined operation of the solar PV and grid is more than load demand, then excess energy is stored in the battery storage systems, which are supplied whenever required to meet the energy deficit. If the battery storage system is unable to meet the load, then the main grid (NITD-DVC-grid) supplies energy to meet the total load demand. The diesel generator is used to meet the total load demand under an outage or emergency condition to maintain reliable and resilient operation of the system. The proposed grid-connected microgrid is implemented with the combined platform of HOMER-grid and MATLAB; the main objective is to minimize the per-unit cost of energy and net present cost. The flow chart of the proposed system design methodology is shown in Figure 7.5.

The HOMER-grid software created by the National Renewable Energy Laboratory is employed to design and optimize the grid-connected microgrid solution in the projected case study. HOMER-grid covers the simulation and design with probable

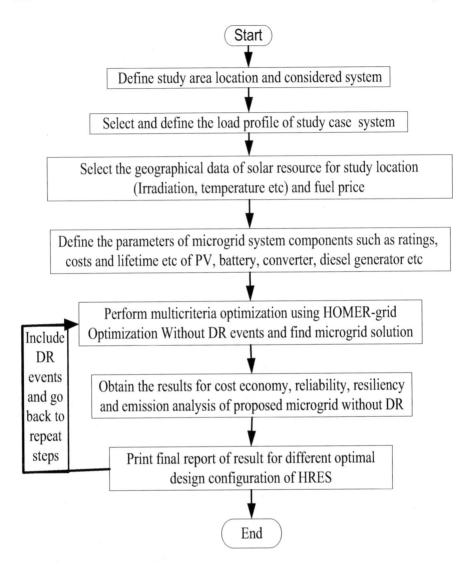

FIGURE 7.5 Flow chart for design methodology of proposed microgrid system.

constraints of the system for optimization. HOMER-grid is innovative program design software to construct the new model and its operation for the renewable energy system-based microgrid design with a grid-connected system. The system simulation and analysis fundamentally depend on the choice of system components selected in any study. HOMER-grid allows the development of system components in huge amounts and sizing the for the successful operations of entire system and also evaluates the total amount of solar PV, battery and converter in the study. The optimum dynamic system design is selected during the grid-connected microgrid

Utility Tariff Variation and Demand Response Events 147

planning through HOMER-grid optimization and simulation analysis based on the minimum net present cost and energy per unit cost. The reduction of demand charge and other cost-saving benefits can be achieved by different options for reducing electricity bills.

The comparison of different costs and savings for installing different combinations of batteries, solar panels and generators are also presented. HOMER-grid uses powerful optimization to find the system's minimum cost of energy with a minimum net present cost that will maximize the savings of the system. A bill from an electric utility can be composed of different types of charges. The charge for energy is the total charge for consuming the quantity of energy in kWh for the month. The demand charge is generally the charge for the highest peak power draw in kW or megawatts (MW) for the month. Also, the fixed charge is a charge that is the same every month and is not affected by consumption or peak demand. Here, the optimization tool considers generators as a system for peak shaving and demand charge reduction. The total cost, capital cost, O&M cost, replacement cost, fuel cost and other parameters are also obtained with HOMER-grid for the hybrid system.

The demand response program is used by the utility to shift or lessen electricity consumption during peak load hours, which offers incentives to customers with different time-of-use rates; also, during peak load events, these DR programs can permit the utility to avoid new investment for energy generation. HOMER-grid computes the demand limit first, including the demand charge, which produces a separate demand limit for all demand response events. It then executes optimization for demand response events to minimize overall costs. If multiple demand response events are included for a month, then chronological order is used in the optimization process. Finally, HOMER-grid optimizes all the configurations of the grid-connected microgrid system based on the multicriteria optimization approach based on various economic and technical parameters such minimization of total net present cost, levelized cost of energy, total annualized cost, reliability, emissions, maximization of reliability and resiliency and so on. Therefore, the different technical and economic performance parameters for the proposed grid-connected microgrid are defined as the following.

7.4.1 Reliability of System

Reliability is the ability of the system to maintain operations during the outage of the utility for a short time, generally lasting a few hours or less. These outages may occur multiple times in a year. Maintaining operations during longer outages comes under resiliency. Systems with energy storage or backup generators can continue operations under the grid outage condition. In the proposed HRES system, reliability is simulated based on outages that happen 10 times per year on average for a 2-hour average repair time. No electrical energy is purchased or sold by the grid during the outage time steps. Also, the reliability of the system is evaluated by the concept of loss of power supply probability. The LPSP can be defined as the probability of deficient power supply consequences when the renewable hybrid system cannot fulfill the load demand. The LPSP can be evaluated based on the following equation [8]:

$$LPSP = \frac{\sum_{t=0}^{T} P \times Failure \times Time\, with\left(P^t_{Supplied} < P^t_{demand}\right)}{N_t} \quad (7.7)$$

where T is the entire period and N_t is the intervals for deficit energy.

7.4.2 Resilience of System

Resilience can be defined as the ability of system for a facility to respond to an extended, multi-day utility outage like a natural disaster such as hurricanes or wildfires. It provides a supplementary value stream by offering a hybrid/generator project's value in extended outages. In the proposed HRES system, resilience is simulated based on an outage that occurs once in every 1.0 years. The simulated grid shutdown starts on 21 September 03:00 and ends on 23 September 03:00 after 2 days. During the resilience outage duration, the HRES work as stand-alone systems (i.e., disconnected from grid) for resiliency applications. During a resilience event, only the critical/primary electrical load and thermal load are served, not non-critical loads.

7.4.3 Demand Response

The demand response model for the proposed study is used during peak load hours to shift or reduce energy consumption, which offers economic incentives. In this case study, the demand reduction incentives are considered to be 35 ₹/kW for all DR events. Demand response programs are incentives offered by utilities to lower consumption at certain times during the year. If the system can manage the facilities to reduce demand during the event successfully, the utility would pay the pre-approved amount for every kW reduced, which is known as the demand reduction incentive. The NIT Durgapur utility (DVC-NITD) tariff offers an incentive of ₹ 35.00 for every kW reduced. So, it's required to reduce consumption for a period of 3.0 hours when notified by the utility. The start-up for this program leads to total revenue for the proposed grid-connected microgrid system. Six demand response events occur in a year for the proposed case study, as given in the following, during which the revenue is incurred.

- Event-1: 15 Mar 03:00 PM
- Event-2: 16 Apr 02:00 PM
- Event-3: 21 Sep 04:00 PM
- Event-4: 08 Oct 03:00 PM
- Event-5: 27 Nov 02:00 PM
- Event-6: 24 Dec 01:00 PM

7.4.4 Total Net Present Cost

The total net present cost (TNPC) of a system can be defined as the total value of the present system cost components, including initial cost, replacement cost, operation and maintenance cost, emissions penalty, fuel costs and power purchasing costs grid over the project lifetime. The total net present cost of a system is the main economic parameter. The different system configurations are ranked, which is used to estimate the levelized cost of energy and total annualized cost.

7.4.5 Levelized Cost of Energy

The levelized cost of energy (LCOE) is an eminent noteworthy factor to determine cost-effective analysis of a distribution grid or network in the presence of renewable sources. The average cost of energy or LCOE can be defined as [9–10]:

$$LCOE = \frac{C_{TNPC}}{\sum_{H=1}^{H=8760} P_{demand}} \times CRF \qquad (7.8)$$

where P_{demand} is energy consumed (kWh) hourly, C_{TNPC} is total net present cost and capital recovery factor is denoted as CRF, which is determined by the following equation:

$$CRF = \frac{r(1+r)^n}{(1+r)^n - 1} \qquad (7.9)$$

where r is the actual rate of interest, and n is system/project lifetime.

7.4.6 Simple Payback

Simple payback is the number of years in which cumulative cash flow changes move towards positive values from the proposed system's negative values relative to system base case. The payback indicates how much time it will take to recover the investment difference costs between the base case system and the proposed system.

7.4.7 Pollutant Emissions and Particulate Matter

Pollutant emissions (CO_2, NO_x, SO_2, etc.) and particulate matter are taken into consideration as social factors. The grid is majorly supplied with thermal power plants; therefore, generally the pollutant emission component consists of CO (carbon monoxide), CO_2 (carbon dioxide), SO_2 (sulfur dioxide), NO_x (nitrogen oxides) and particulate matter. These pollutant emission elements badly affect the environment as well as human health. Mainly, PM affects human health adversely by affecting the respiratory system and other health issues. The equivalent PM consists of $M_{2.5}$ (air pollutant particles less than or equal to 2.5 μm diameter) and PM_{10} (air pollutant particles less than or equal to 10.0 μm diameter). The pollutant emission factor for CO_2 is 632.0 g/kWh, SO_2 is 2.74 g/kWh and NO_x is 1.34 g/kWh [17]. Also, particulate matter is given on the basis of a linear regression equation with correlation $r=0.884$ of PM and carbon weight [17]:

$$PM = (0.47 \times W_{carbon}) + 0.12 \quad \text{with correlation } r = 0.884 \qquad (7.10)$$

where $PM = PM_{2.5} + PM_{10}$, and W_{carbon} is the weight of total carbon emissions.

7.5 RESULTS AND DISCUSSION

The results of the analysis of the proposed microgrid design comprising solar PV, battery storage and a diesel generator connected with the DVC-NITD-grid (main grid utility) by using HOMER-grid multicriteria optimization to fulfill the load demand of the study area is presented with demand response events for the two optional tariffs of the grid utility. The feasible design and sizing of the proposed microgrid and parameter estimation are performed on the basis of minimum net present cost and cost of energy with simultaneous maximization of reliability for available renewable resources in the NITD system case study. The best configurations are analyzed and discussed based on their technical and economic performance benefits using HOMER-grid multicriteria optimization in the design and analysis of various configurations. The optimal design solution of the proposed grid-connected microgrid is chosen based on the higher feasibility of the proposed utility grid-connected microgrid. A detailed analysis of implemented system cases is presented for two different tariff options for the utility grid, including demand response events. The first is the base case system or current system of NITD, which includes only the diesel generator connected with the DVC-NITD-grid, and the second system is the proposed microgrid consisting of solar PV, battery storage and a diesel generator connected to the DVC-NITD-grid (i.e., the main utility grid). The impact of both tariffs and demand response is analyzed for these systems.

7.5.1 BASE CASE SYSTEM (CURRENT EXISTING SYSTEM OF NITD): DIESEL GENERATOR CONNECTED TO DVC-NITD-GRID

The base system or current system is the existing grid-connected (DVC-NITD-grid) system of NITD with a diesel generator. The average electric energy needs the of current existing system at NITD, WB, India, are 11,687.2 kWh/day, 355.5 MWh/month and 4,265.8 MWh/year, with a peak hourly demand of 1,488.52 kW. The load profile of the base case system for the day on which the largest demand occurs and the energy consumption patterns are shown in Figures 7.6 and 7.7, respectively. It is currently connected to the DVC-NITD-grid with a tariff-1 scheme and diesel generator units of 1,500 kW total capacity. The two optional tariff schemes of the grid utility, tariff scheme-1 and tariff scheme-2 for the case study, are shown in Figure 7.8.

NITD is currently spending ₹ 23.99M on the utility bill per year with tariff-1, in which 24% of the utility bill is demand charges, as given in Table 7.4. The total electric consumption of 11,687.2 kWh/day and peak of 1,488.52 kW electrical load in the base case system is served by two energy supply sources, including outage conditions without DR, which is shown in Table 7.3. The annual energy supply is 42,83,279 kWh/yr to fulfill the load demand of 42,64,777 kWh/yr for the base case system without DR. The monthly energy consumption served by sources including both tariffs under DR events for base system is represented in Figure 7.9. The total annual load is reduced to 42,60,657 kWh/yr with DR events; therefore, the energy purchased by the grid utility also decreases for the base case system. The monthly utility charge detail with tariff-1 and tariff-2 for the base case system without and with DR events are given in Tables 7.4 and 7.5, respectively, which include total energy purchased from the grid, energy sold to the grid, net energy purchased, energy charges and demand charges of the utility grid.

Utility Tariff Variation and Demand Response Events 151

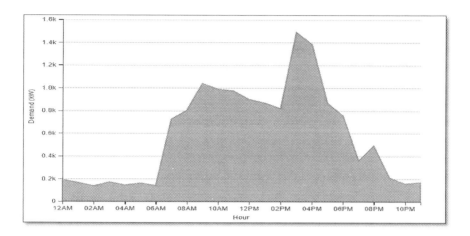

FIGURE 7.6 Load profile for the day on which peak demand occurs.

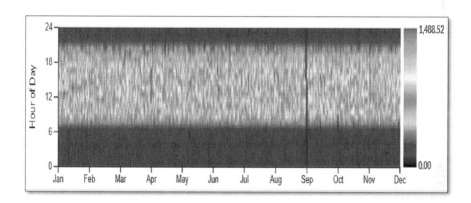

FIGURE 7.7 Energy consumption pattern.

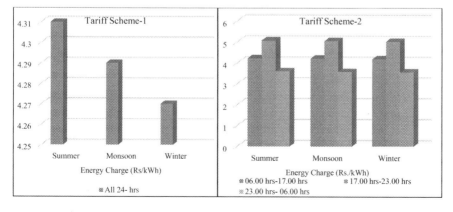

FIGURE 7.8 Two option tariffs for the grid utility, tariff scheme-1 and tariff scheme-2 for the case study.

TABLE 7.3
Annual Base Case System Energy Supply and Consumption for Both Tariffs with DR Events

Without DR Events					With DR Events				
Annual Energy Supply with Both Tariffs			Annual Energy Consumption with Both Tariffs		Annual Energy Supply with Both Tariffs for DR Events			Reduced Annual Energy Consumption with Both Tariffs for DR Events	
Energy source	kWh/yr	%	Total annual load kWh/yr	42,64,777	Energy source	kWh/yr	%	Total annual load kWh/yr	42,60,657
DG set	31,125	0.727	Excess energy kWh/yr	18,502	DG set	24,750	0.579	Excess energy kWh/yr	15,414
Grid Purchases	42,52,154	99.3	Total	42,83,279	Grid Purchases	42,51,321	99.4	Total	42,76,070
Total	42,83,279	100			Total	42,76,071	100		

The total charge without DR events for tariff-1 is ₹ 2,39,99,581.82 and for tariff-2 is ₹ 2,42,61,542.36, which are reduced by implementing DR events to ₹ 2,38,08,287.35 for tariff-1 and ₹ 2,40,71,345.90 for tariff-2. The net present and annual cost of the base case system without DR events are ₹ 31,78,10,221.04 and ₹ 2,45,84,011.93 for tariff-1 and ₹ 32,11,96,720.18 and ₹ 2,48,45,972.46 for tariff-2, as specified in Tables 7.6 and 7.7. Also, the net present and annual cost of base case system with DR events are ₹ 31,50,49,871.62 and ₹ 2,43,70,486.82 for tariff-1 and ₹ 31,84,50,565.45 and ₹ 2,46,33,545.37 for tariff-2.

The comparative summary for COE, net present cost, operating cost and savings of the base system with respect to DR events, including different tariffs, are shown in Table 7.8, which clearly indicates that the minimum COE, net present cost and operating cost are obtained for tariff-1 under DR events. The relative comparison of COE, operating and net present cost of the base system for both tariffs with DR events are represented in Figure 7.10. The DG operational parameters for both tariffs with and without DR events in the base system are given in Table 7.9.

The DG operation and power output profile for both tariffs and DR events under outage for reliability (short outage) and resilience (long outage) for the base system are demonstrated in Table 7.10 and Figure 7.11 separately, which show they operate to maintain the reliability and resilience operation of the system under short and long outage conditions. The long outages correspond to whole-day or multiple-day utility outages for natural disasters such as hurricanes or wildfires. It provides an additional value stream by proving the value of hybrid power generation

TABLE 7.4
Utility Charges with tariff-1 and tariff-2 for Base Case System without DR Events

Month	Energy Purchased for Both Tariffs (kWh)	Energy Sold for Both Tariffs (kWh)	Net Energy Purchased for Both Tariffs (kWh)	Demand Charge for Both Tariffs	Energy Charge for tariff-1	Energy Charge for tariff-2	Total for tariff-1	Total for tariff-2
Jan	3,56,052	0	3,56,052	₹4,60,536.53	₹15,20,339.92	₹15,44,148.80	₹19,80,876.45	₹20,04,685.33
Feb	3,18,927	0	3,18,927	₹5,10,848.02	₹13,61,818.45	₹13,82,464.96	₹18,72,666.47	₹18,93,312.98
Mar	3,76,548	0	3,76,548	₹5,06,099.99	₹16,22,921.95	₹16,46,111.79	₹21,29,021.94	₹21,52,211.78
Apr	3,54,091	0	3,54,091	₹4,79,816.37	₹15,26,130.14	₹15,48,801.67	₹20,05,946.51	₹20,28,618.04
May	3,57,403	0	3,57,403	₹4,38,159.12	₹15,40,406.36	₹15,64,459.69	₹19,78,565.48	₹20,02,618.82
Jun	3,54,219	0	3,54,219	₹4,52,765.52	₹15,26,684.76	₹15,48,102.06	₹19,79,450.28	₹20,00,867.58
Jul	3,59,643	0	3,59,643	₹5,01,718.25	₹15,42,866.85	₹15,64,449.51	₹20,44,585.10	₹20,66,167.76
Aug	3,77,332	0	3,77,332	₹5,29,250.92	₹16,18,756.09	₹16,40,986.85	₹21,48,007.00	₹21,70,237.77
Sep	3,34,566	0	3,34,566	₹4,87,801.09	₹14,35,287.49	₹14,56,259.98	₹19,23,088.58	₹19,44,061.07
Oct	3,59,390	0	3,59,390	₹4,62,379.75	₹15,41,784.26	₹15,59,780.98	₹20,04,164.01	₹20,22,160.73
Nov	3,47,543	0	3,47,543	₹4,22,547.23	₹14,84,007.01	₹15,07,303.75	₹19,06,554.24	₹19,29,850.98
Dec	3,64,131	0	3,64,131	₹4,71,814.46	₹15,54,841.30	₹15,74,935.07	₹20,26,655.76	₹20,46,749.52
Total Annual	42,59,844	0	42,59,844	₹57,23,737.24	₹1,82,75,844.59	₹1,85,37,805.12	₹2,39,99,581.82	₹2,42,61,542.36

TABLE 7.5
Utility Charges with Different Tariffs for Base Case System with DR Events

Month	Energy Purchased for Both Tariffs (kWh)	Energy Sold for Both Tariffs (kWh)	Net Energy Purchased for Both Tariffs (kWh)	Demand Charge for Both Tariffs	Energy Charge for tariff-1	Energy Charge for tariff-2	Total for tariff-1	Total for tariff-2
Jan	3,58,462	0	3,58,462	₹ 4,60,536.53	₹ 15,30,632.88	₹ 15,54,967.24	₹ 19,91,169.42	₹ 20,15,503.77
Feb	3,17,236	0	3,17,236	₹ 5,10,848.02	₹ 13,54,597.99	₹ 13,75,486.52	₹ 18,65,446.02	₹ 18,86,334.55
Mar	3,74,528	0	3,74,528	₹ 5,06,099.99	₹ 16,14,215.10	₹ 16,36,899.53	₹ 20,92,849.62	₹ 21,15,534.05
Apr	3,50,218	0	3,50,218	₹ 4,79,816.37	₹ 15,09,439.70	₹ 15,32,199.37	₹ 19,60,243.31	₹ 19,83,002.99
May	3,57,403	0	3,57,403	₹ 4,38,159.12	₹ 15,40,405.09	₹ 15,64,458.64	₹ 19,78,564.22	₹ 20,02,617.76
Jun	3,55,007	0	3,55,007	₹ 4,52,765.52	₹ 15,30,078.46	₹ 15,49,650.68	₹ 19,82,843.98	₹ 20,02,416.21
Jul	3,60,794	0	3,60,794	₹ 5,01,718.25	₹ 15,47,807.62	₹ 15,70,327.10	₹ 20,49,525.86	₹ 20,72,045.34
Aug	3,77,017	0	3,77,017	₹ 5,29,250.92	₹ 16,17,404.14	₹ 16,40,411.67	₹ 21,17,775.09	₹ 21,40,782.62
Sep	3,34,707	0	3,34,707	₹ 4,87,801.09	₹ 14,35,894.65	₹ 14,56,855.82	₹ 19,23,695.74	₹ 19,44,656.91
Oct	3,58,765	0	3,58,765	₹ 4,62,379.75	₹ 15,39,101.18	₹ 15,57,629.06	₹ 19,78,890.59	₹ 19,97,418.47
Nov	3,44,877	0	3,44,877	₹ 4,22,547.23	₹ 14,72,622.81	₹ 14,96,132.83	₹ 18,70,746.95	₹ 18,94,256.97
Dec	3,62,307	0	3,62,307	₹ 4,71,814.46	₹ 15,47,051.30	₹ 15,67,291.01	₹ 19,96,536.55	₹ 20,16,776.26
Total Annual	42,51,321	0	42,51,321	₹ 57,23,737.24	₹ 1,82,39,250.93	₹ 1,85,02,309.49	₹ 2,38,08,287.35	₹ 2,40,71,345.90

TABLE 7.6
Net Present and Annual Cost with tariff-1 and tariff-2 for Base System without DR Events

Cost Components	DG set for Both Tariffs	Utility Charges-1	Utility Charges-2	Total System with tariff-1	Total System with tariff-2
Net Present Cost of System					
Capital	₹ 11,25,000.00	₹ 0.00	₹ 0.00	₹ 11,25,000.00	₹ 11,25,000.00
Replacement	₹ 0.00	₹ 0.00	₹ 0.00	₹ 0.00	₹ 0.00
O&M	₹ 13,379.98	₹ 31,02,54,991.19	₹ 31,36,41,490.33	₹ 31,02,68,371.17	₹ 31,36,54,870.31
Fuel	₹ 65,75,856.39	₹ 0.00	₹ 0.00	₹ 65,75,856.39	₹ 65,75,856.39
Salvage	₹ -1,59,006.52	₹ 0.00	₹ 0.00	₹ -1,59,006.52	₹ -1,59,006.52
Total	₹ 75,55,229.85	₹ 31,02,54,991.19	₹ 31,36,41,490.33	₹ 31,78,10,221.04	₹ 32,11,96,720.18
Annual Cost of System					
Capital	₹ 87,023.68	₹ 0.00	₹ 0.00	₹ 87,023.68	₹ 87,023.68
Replacement	₹ 0.00	₹ 0.00	₹ 0.00	₹ 0.00	₹ 0.00
O&M	₹ 1,035.00	₹ 2,39,99,581.82	₹ 2,42,61,542.36	₹ 2,40,00,616.82	₹ 2,42,62,577.36
Fuel	₹ 5,08,671.28	₹ 0.00	₹ 0.00	₹ 5,08,671.28	₹ 5,08,671.28
Salvage	₹ -12,299.85	₹ 0.00	₹ 0.00	₹ -12,299.85	₹ -12,299.85
Total	₹ 5,84,430.10	₹ 2,39,99,581.82	₹ 2,42,61,542.36	₹ 2,45,84,011.93	₹ 2,48,45,972.46

TABLE 7.7
Net Present and Annual Cost with Different Tariffs for Base System with DR Events

Cost Components	DG set for Both Tariffs	Utility Charges-1	Utility Charges-2	Total System with tariff-1	Total System with tariff-2
Net Present Cost of System					
Capital	₹ 11,25,000.00	₹ 0.00	₹ 0.00	₹ 11,25,000.00	₹ 11,25,000.00
Replacement	₹ 0.00	₹ 0.00	₹ 0.00	₹ 0.00	₹ 0.00
O&M	₹ 12,798.24	₹ 30,77,82,028.65	₹ 31,11,82,722.48	₹ 30,77,94,826.89	₹ 31,11,95,520.72
Fuel	₹ 62,89,949.59	₹ 0.00	₹ 0.00	₹ 62,89,949.59	₹ 62,89,949.59
Salvage	₹ -1,59,904.86	₹ 0.00	₹ 0.00	₹ -1,59,904.86	₹ -1,59,904.86
Total	₹ 72,67,842.97	₹ 30,77,82,028.65	₹ 31,11,82,722.48	₹ 31,50,49,871.62	₹ 31,84,50,565.45
Annual Cost of System					
Capital	₹ 87,023.68	₹ 0.00	₹ 0.00	₹ 87,023.68	₹ 87,023.68
Replacement	₹ 0.00	₹ 0.00	₹ 0.00	₹ 0.00	₹ 0.00
O&M	₹ 990.00	₹ 2,38,08,287.35	₹ 2,40,71,345.90	₹ 2,38,09,277.35	₹ 2,40,72,335.90
Fuel	₹ 4,86,555.14	₹ 0.00	₹ 0.00	₹ 4,86,555.14	₹ 4,86,555.14
Salvage	₹ -12,369.34	₹ 0.00	₹ 0.00	₹ -12,369.34	₹ -12,369.34
Total	₹ 5,62,199.47	₹ 2,38,08,287.35	₹ 2,40,71,345.90	₹ 2,43,70,486.82	₹ 2,46,33,545.37

TABLE 7.8
Comparative Summary of Base System for Both Tariffs with DR Events

Parameters	Without DR Events		With DR Events	
	Tariff-1	Tariff-2	Tariff-1	Tariff-2
COE (₹/kWh)	5.76 ₹/kWh	5.82 ₹/kWh	5.72 ₹/kWh	5.78 ₹/kWh
Net present cost (₹)	₹ 317810200	₹ 321196700	₹ 315049900	₹ 318450600
Operating cost (₹)	₹ 24496990	₹ 24758950	₹ 24283460	₹ 24546520
Utility bill savings (₹)	NA (Base case)	₹ -261962	NA (Base case)	₹ -263060
Total bill savings (₹)	NA (Base case)	₹ -3386518	NA (Base case)	₹ -3400713

TABLE 7.9
DG Set Operational Parameters for Both Tariffs with DR Event Effects in Base System

Parameters	Value for Both Tariffs without DR Events	Value for Both Tariffs with DR Events
Hours of Operation	83.0 hrs/yr	66.0 hrs/yr
Number of Starts	11.0 starts/yr	11.0 starts/yr
Operational Life	181 yr	227 yr
Electrical Production	31,125 kWh/yr	24,750 kWh/yr
Mean Electrical Output	375 kW	375 kW
Fuel Consumption	9,271 L	7,372 L/yr
Generator Fuel Price	66.0 ₹/L	66.0 ₹/L

TABLE 7.10
DG Operation for Both Tariffs and DR Events under Outage for Reliability (Short Outage) and Resilience (Long Outage) for Base System

Components	Value
DG runtime (hours/day)	24.0
DG O&M cost (₹/day)	360
DG fuel consumption (L/day)	2,681
DG fuel cost (₹/day)	1,76,929

during outages, which are considered for resilience analysis. The shorter outages are more frequent outages defined for the reliability analysis. In this case study, along outage is defined and simulated that occurs once in every 1.0 years. It is assumed that the grid shutdown starts on 21 September 03:00 and ends on 23

TABLE 7.11
Pollutant Emissions of the Base Case System for Both Tariffs with DR Events

Pollutant Emission	Emission Amount for Both Tariffs without DR Events	Emission Amount for Both Tariffs with DR Events
Carbon Dioxide	27,11,629 kg/yr	27,06,132 kg/yr
Carbon Monoxide	153 kg/yr	122 kg/yr
Unburned Hydrocarbons	6.68 kg/yr	5.31 kg/yr
Particulate Matter	0.927 kg/yr	0.737 kg/yr
Sulfur Dioxide	11,710 kg/yr	11,696 kg/yr
Nitrogen Oxides	5,842 kg/yr	5,811 kg/yr

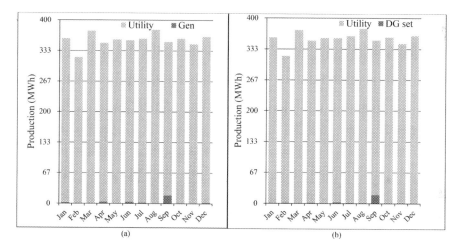

FIGURE 7.9 Monthly energy consumption served by sources including both tariffs for base system (a) without DR events, (b) with DR events.

September 03:00. Also, short outages occur 10 times per year on average with a 2.0 hour average repair time. The system performance during these outages are shown in Figure 7.11(b) and (c). The pollutant emissions for the base case system considering the different tariffs under DR events are reported in Table 7.11. The emissions of these pollutants are considered from the production of electricity by generators and consumption of grid electricity. The pollutant emissions of the base system such as carbon dioxide without DR and with DR events are 27,11,629 kg/yr and 27,06,132 kg/yr, and the particulate matter without DR and with DR events are 0.927 kg/yr and 0.737 kg/yr.

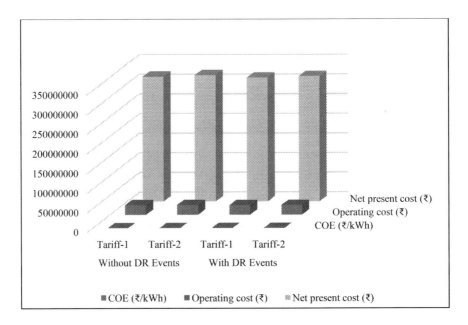

FIGURE 7.10 COE, operating and net present cost of base system for both tariffs with DR events.

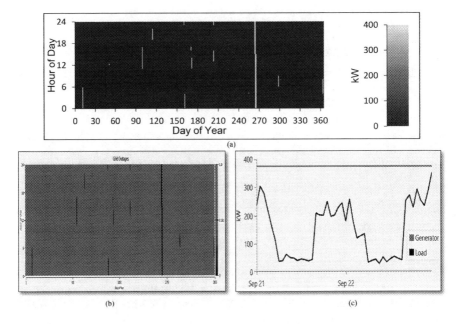

FIGURE 7.11 (a) Power output pattern of DG under outage, (b) grid outage and reliability (short outage) and resilience (long outage), (c) system operation under (long outage under resilience) for base system including DR events for both tariffs.

7.5.2 Proposed Microgrid System: Solar PV-Battery Storage-Diesel Generator Connected to Main Utility Grid

The proposed microgrid system includes the installation of such components as solar PV, wind battery storage, a system converter and diesel generator connected with the DVC-NITD-grid system, which is obtained as a feasible solution based on the minimum values of the cost of energy and net present cost. This microgrid system proposes adding 500 kW of solar PV, 4,902 kWh/5,515 kWh of battery capacity and 1,042 kW/977 kW system converter with the base case system for different tariffs, as given in Table 7.12. The proposed microgrid system reduces the system annual utility bill to ₹ 18.1M for tariff-1 and ₹ 17.68M for tariff-2 without DR events; with DR events, the bills are ₹ 17.83M for tariff-1 and ₹ 17.44M tariff-2.

Solar PV: The solar PV system's detailed operational parameters for the proposed microgrid are given in Table 7.13, which specifies that the solar PV has a nominal

TABLE 7.12

Component Installation Detail for the Proposed Microgrid for Different Tariffs, Including DR Events Obtained by HOMER-grid

Components	Tariff-1 Size	Tariff-1 Installation/Capital Cost	Tariff-2 Size	Tariff-2 Total Installation/Capital Cost
Solar PV	500 kW	₹ 10,69,394	500 kW	₹ 10,69,394
Battery Storage	4,902 kWh	₹ 26,96,100	5,515 kWh	₹ 30,33,250
System Converter	1,042 kW	3,12,600	977 kW	₹ 2,93,062.5
DG Set	1,500 kW	₹ 11,25,000.00	1,500 kW	₹ 11,25,000.00

TABLE 7.13

Solar PV Details for Both Tariffs and DR Events

Parameters	Values
Rated capacity	500 kW
Mean output	87.1 kW
Mean output per day	2,090 kWh/d
Capacity factor	17.4%
Total production	7,62,863 kWh/yr
Minimum output	0 kW
Maximum output	498 kW
PV penetration	17.8%
Hours of operation	4,390 hrs/yr
PV levelized cost	0.11 ₹/kWh

capacity of 500 kW with the annual production of 7,62,863 kWh/yr. The yearly operating hours are 4,390 hrs/yr, with a per-day mean output of 2,090 kWh/day.

Battery Storage System: The battery storage system has a nominal capacity of 4,902 kWh/5,515 kWh and specific annual throughput considering different tariffs with DR events. The detailed operational parameters of battery storage for the proposed microgrid system are specified in Table 7.14.

Diesel Generator Set: The diesel generator set has a capacity of 1,500 kW which provides different annual power outputs during operation, including different operation hours for different tariffs with and without DR events. The operational parameter summary of DG for the proposed microgrid system is given in Table 7.15.

System Converter: The system converter has two operating modes, inverter mode and rectifier mode, with a nominal capacity of 1,042 kW/ 977 kW for different tariffs with and without DR events. The detailed operational parameters of the system converter for the proposed microgrid system are given in Table 7.16.

The total electric energy supplied by distinct sources like the solar PV, DG set and utility grid considered without DR events are 44,34,765 kWh/yr for tariff-1 and 45,12,505 kWh/yr for tariff-2, and with DR events, they are 44,24,594 kWh/yr for tariff-1 and 45,06,388 kWh/yr for tariff-2, as illustrated in Table 7.17. The monthly energy consumption served by energy sources solar PV, DG and grid for the proposed microgrid considered without and with DR events is represented in Figure 7.12. The supplied energy is mainly utilized to fulfill the load demand and to charge the batteries; also, a portion is used to support system losses. The total monthly grid utility charges for the proposed grid-connected solar PV-based microgrid system without DR events are ₹ 1,80,90,790.08 for tariff-1 and ₹ 1,76,80,120.03 for tariff-2 per year, as illustrated in Tables 7.18 and 7.19, respectively. Similarly, the total grid utility charges for the proposed microgrid system with DR events are ₹ 1,78,32,177.29 for tariff-1 and ₹ 1,74,42,214.90 for tariff-2 annually, as shown in Tables 7.20 and 7.21 individually.

TABLE 7.14
Battery Storage Operational Details for Both Tariffs and DR Events

Parameters	Unit	Without DR		With DR	
		Tariff-1	Tariff-2	Tariff-1	Tariff-2
Rated capacity	kWh	4,902	5,515	4,902	5,515
Maintenance cost	₹/yr	49,020	55,150	49,020	55,150
Autonomy	hr	8	9	8	8
Annual throughput	kWh/yr	6,39,811	9,98,297	6,36,683	10,09,627
Energy in	kWh/yr	6,74,420	10,51,775	6,71,123	10,63,671
Energy out	kWh/yr	6,06,978	9,47,068	6,04,010	9,57,816
Battery wear cost	₹/kWh	0.193	0.193	0.193	0.193
Losses	kWh/yr	62,540	99,688	62,210	1,01,493
Storage depletion	kWh/yr	−4,902	−5,020	−4,902	−4,362

TABLE 7.15
Diesel Generator Operational Details

Parameters	Unit	Without DR Tariff-1	Without DR Tariff-2	With DR Tariff-1	With DR Tariff-2
Hours of operation	hrs/yr	5	8	5	6
Number of starts	starts/yr	1	3	1	2
Operational life	yr	3,000	1,875	3,000	2,500
Capacity factor	%	0.0314	0.0448	0.0314	0.038
Fixed generation cost	₹/hr	1,597	1,597	1,597	1,597
Marginal generation cost	₹/kWh	15.6	15.6	15.6	15.6
Electrical production	kWh/yr	4,126	5,883	4,126	4,995
Mean electrical output	kW	825	735	825	833
Minimum electrical output	kW	404	375	404	392
Maximum electrical output	kW	1,369	1,333	1,369	1,141
Fuel consumption	L/yr	1,090	1,574	1,090	1,318
Specific fuel consumption	L/kWh	0.264	0.268	0.264	0.264
Fuel energy input	kWh/yr	10,723	15,487	10,723	12,969
Mean electrical efficiency	%	38.5	38	38.5	38.5

TABLE 7.16
System Converter Operational Details

Parameters	Unit	Without DR Tariff-1 Inverter Mode	Without DR Tariff-1 Rectifier Mode	Without DR Tariff-2 Inverter Mode	Without DR Tariff-2 Rectifier Mode	With DR Tariff-1 Inverter Mode	With DR Tariff-1 Rectifier Mode	With DR Tariff-2 Inverter Mode	With DR Tariff-2 Rectifier Mode
Operational capacity	kW	1,042	1,042	977	977	1,042	1,042	977	977
Mean output	kW	141	69.4	182	116	141	69	184	119
Minimum output	kW	0	0	0	0	0	0	0	0
Maximum output	kW	963	1,042	977	977	963	1,042	977	977
Capacity factor	%	14	6.66	18.6	11.9	14	6.62	18.9	12.2
Hours of operation	hrs/yr	4,866	2,287	5,062	2,932	4,871	2,274	5,212	2,769
Energy out	kWh/yr	12,37,823	6,07,585	15,94,670	10,20,444	12,35,074	6,04,340	16,14,834	10,42,817
Energy in	kWh/yr	13,02,972	6,39,563	16,78,600	10,74,152	13,00,078	6,36,147	16,99,825	10,97,702
Losses	kWh/yr	65,149	31,978	83,930	53,708	65,004	31,807	84,991	54,885

TABLE 7.17
Annual Energy Supply Summary of Proposed Microgrid for Different Tariffs and DR Events

	Without DR				With DR			
	Tariff-1		Tariff-2		Tariff-1		Tariff-2	
Energy Source	kWh/yr	%	kWh/yr	%	kWh/yr	%	kWh/yr	%
Solar PV	7,62,863	17.2	7,62,863	16.9	7,62,863	17.2	7,62,863	16.19
DG Set	4,126	0.0930	5,883	0.130	4,126	0.0933	4,995	0.111
Grid Purchases	36,67,775	82.7	37,43,759	83.0	36,57,604	82.7	37,38,529	83.0
Total	44,34,765	100	45,12,505	100	44,24,594	100	45,06,388	100

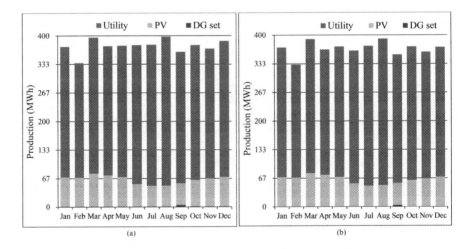

FIGURE 7.12 Monthly energy consumption served by sources for proposed microgrid: (a) without DR events, (b) with DR events.

The total utility charges comprise the total energy purchased from and sold to the grid, net energy purchased, energy charges and demand charges of the utility grid. The net present and annual cost for the proposed microgrid system without DR events are ₹ 24,15,05,393.47 and ₹ 1,86,81,499.46 for tariff-1 and ₹ 23,71,30,140.96 and ₹ 1,83,43,054.53 for tariff-2, respectively, as listed in Tables 7.22 and 7.23. In the same way, the net present and annual cost for the proposed microgrid system with DR events are ₹ 23,81,62,172.41 and ₹ 1,84,22,886.68 for tariff-1 and ₹ 23,33,92,753.81 and ₹1,80,53,951.27 for tariff-2, as enumerated in Tables 7.24 and 7.25. The total net present and annual cost of the system consist of capital cost, replacement cost, O&M cost, fuel cost and salvage cost.

TABLE 7.18
Utility Charges with tariff-1 for Proposed Microgrid without DR Events

Month	Energy Purchased (kWh)	Energy Sold (kWh)	Net Energy Purchased (kWh)	Energy Charge	Demand Charge	Total Charges
Jan	3,00,693	0	3,00,693	₹ 12,83,960.59	₹ 1,87,332.83	₹ 14,71,293.43
Feb	2,63,992	0	2,63,992	₹ 11,27,246.43	₹ 1,68,287.17	₹ 12,95,533.61
Mar	3,14,426	0	3,14,426	₹ 13,55,174.17	₹ 1,74,334.31	₹ 15,29,508.48
Apr	2,95,886	0	2,95,886	₹ 12,75,266.80	₹ 1,92,426.35	₹ 14,67,693.14
May	3,03,882	1.16	3,03,881	₹ 13,09,727.73	₹ 1,71,954.63	₹ 14,81,682.36
Jun	3,08,966	0	3,08,966	₹ 13,31,644.07	₹ 2,46,467.60	₹ 15,78,111.67
Jul	3,24,228	0	3,24,228	₹ 13,90,936.39	₹ 1,96,898.29	₹ 15,87,834.67
Aug	3,42,148	0	3,42,148	₹ 14,67,815.89	₹ 2,06,187.33	₹ 16,74,003.21
Sep	2,99,526	0	2,99,526	₹ 12,84,966.15	₹ 2,15,191.43	₹ 15,00,157.58
Oct	3,12,638	0	3,12,638	₹ 13,41,216.74	₹ 1,73,512.81	₹ 15,14,729.55
Nov	2,97,317	0	2,97,317	₹ 12,69,545.56	₹ 1,96,088.32	₹ 14,65,633.88
Dec	3,04,073	0	3,04,073	₹ 12,98,392.49	₹ 2,26,215.99	₹ 15,24,608.49
Total Annual	36,67,775	1.16	36,67,774	₹ 1,57,35,893.00	₹ 23,54,897.07	₹ 1,80,90,790.08

TABLE 7.19
Utility Charges with tariff-2 for Proposed Microgrid without DR Events

Month	Energy Purchased (kWh)	Energy Sold (kWh)	Net Energy Purchased (kWh)	Energy Charge	Demand Charge	Total Charges
Jan	3,04,590	0	3,04,590	₹ 12,58,583.36	₹ 1,74,140.38	₹ 14,32,723.73
Feb	2,67,808	0	2,67,808	₹ 11,09,061.71	₹ 1,63,531.23	₹ 12,72,592.94
Mar	3,17,445	0	3,17,445	₹ 13,31,918.34	₹ 1,71,072.34	₹ 15,02,990.68
Apr	3,01,985	0	3,01,985	₹ 12,54,820.07	₹ 1,78,681.62	₹ 14,33,501.69
May	3,07,151	0	3,07,151	₹ 12,84,864.76	₹ 1,70,699.49	₹ 14,55,564.25
Jun	3,23,845	0	3,23,845	₹ 13,00,172.92	₹ 2,33,376.15	₹ 15,33,549.07
Jul	3,29,718	0	3,29,718	₹ 13,60,999.93	₹ 1,91,508.72	₹ 15,52,508.65
Aug	3,47,811	0	3,47,811	₹ 14,35,909.59	₹ 2,00,123.00	₹ 16,36,032.59
Sep	3,07,123	0	3,07,123	₹ 12,60,182.69	₹ 2,01,348.96	₹ 14,61,531.65
Oct	3,14,869	0	3,14,869	₹ 13,17,800.75	₹ 1,64,634.35	₹ 14,82,435.10
Nov	3,03,859	0	3,03,859	₹ 12,47,904.23	₹ 1,82,773.68	₹ 14,30,677.91
Dec	3,17,556	0	3,17,556	₹ 12,72,888.94	₹ 2,13,122.83	₹ 14,86,011.77
Total Annual	37,43,759	0	37,43,759	₹ 1,54,35,107.28	₹ 22,45,012.75	₹ 1,76,80,120.03

TABLE 7.20
Utility Charges with tariff-1 for Proposed Microgrid with DR Events

Month	Energy Purchased (kWh)	Energy Sold (kWh)	Net Energy Purchased (kWh)	Energy Charge	Demand Charge	Total Charges
Jan	3,02,397	0	3,02,397	₹ 12,91,234.87	₹ 1,87,332.83	₹ 14,78,567.71
Feb	2,62,808	0	2,62,808	₹ 11,22,190.48	₹ 1,68,287.17	₹ 12,90,477.66
Mar	3,12,132	0	3,12,132	₹ 13,45,289.15	₹ 1,75,421.64	₹ 14,82,185.57
Apr	2,92,617	0	2,92,617	₹ 12,61,178.95	₹ 1,91,051.89	₹ 14,10,916.83
May	3,03,882	1.16	3,03,881	₹ 13,09,726.93	₹ 1,71,954.63	₹ 14,81,681.57
Jun	3,09,581	0	3,09,581	₹ 13,34,294.91	₹ 2,46,467.60	₹ 15,80,762.51
Jul	3,24,954	0	3,24,954	₹ 13,94,052.05	₹ 1,96,898.29	₹ 15,90,950.33
Aug	3,41,179	0	3,41,179	₹ 14,63,656.28	₹ 2,06,719.03	₹ 16,29,180.07
Sep	2,99,595	0	2,99,595	₹ 12,85,264.27	₹ 2,15,191.43	₹ 15,00,455.70
Oct	3,11,472	0	3,11,472	₹ 13,36,213.21	₹ 1,73,284.41	₹ 14,81,790.26
Nov	2,94,798	0	2,94,798	₹ 12,58,787.42	₹ 1,95,782.25	₹ 14,19,700.51
Dec	3,02,189	0	3,02,189	₹ 12,90,348.68	₹ 2,27,060.70	₹ 14,85,508.57
Total Annual	36,57,604	1.16	36,57,603	₹ 1,56,92,237.21	₹ 23,55,451.87	₹ 1,78,32,177.29

TABLE 7.21
Utility Charges with tariff-2 for Proposed Microgrid with DR Events

Month	Energy Purchased (kWh)	Energy Sold (kWh)	Net Energy Purchased (kWh)	Energy Charge	Demand Charge	Total Charges
Jan	3,07,785	0	3,07,785	₹ 12,59,174.23	₹ 1,87,332.83	₹ 14,46,507.06
Feb	2,66,182	0	2,66,182	₹ 10,98,546.97	₹ 1,68,287.17	₹ 12,66,834.14
Mar	3,15,558	0	3,15,558	₹ 13,16,506.13	₹ 1,76,871.40	₹ 14,54,138.76
Apr	2,98,466	0	2,98,466	₹ 12,26,294.06	₹ 1,91,051.89	₹ 13,76,031.94
May	3,06,794	0	3,06,794	₹ 12,82,568.48	₹ 1,71,954.63	₹ 14,54,523.12
Jun	3,24,050	0	3,24,050	₹ 12,97,106.33	₹ 2,46,467.60	₹ 15,43,573.93
Jul	3,30,875	0	3,30,875	₹ 13,60,849.54	₹ 1,96,898.29	₹ 15,57,747.82
Aug	3,47,279	0	3,47,279	₹ 14,26,285.98	₹ 2,06,719.03	₹ 15,91,809.77
Sep	3,08,543	0	3,08,543	₹ 12,51,752.70	₹ 2,15,191.43	₹ 14,66,944.13
Oct	3,14,614	0	3,14,614	₹ 13,03,614.44	₹ 1,77,486.39	₹ 14,48,826.97
Nov	3,01,813	0	3,01,813	₹ 12,25,526.23	₹ 1,95,782.25	₹ 13,86,439.31
Dec	3,16,569	0	3,16,569	₹ 12,53,678.06	₹ 2,27,060.70	₹ 14,48,837.95
Total Annual	37,38,529	0	37,38,529	₹ 1,53,01,903.14	₹ 23,61,103.61	₹ 1,74,42,214.90

TABLE 7.22
Net Present and Annual Cost with tariff-1 for Proposed Microgrid without DR Events

Cost Components	System	Generic Flat Plate PV	Diesel Generator	Grid Tariff-1	Battery Storage	System Converter
Net Present Cost of System						
Capital	₹ 52,03,093.94	₹ 10,69,393.94	₹ 11,25,000.00	₹ 0.00	₹ 26,96,100.00	₹ 3,12,600.00
Replacement	₹ 12,76,512.67	₹ 0.00	₹ 0.00	₹ 0.00	₹ 11,43,884.67	₹ 1,32,628.00
O&M	₹ 23,45,14,437.42	₹ 10,772.93	₹ 969.56	₹ 23,38,68,988.07	₹ 6,33,706.86	₹ 0.00
Fuel	₹ 9,29,773.15	₹ 0.00	₹ 9,29,773.15	₹ 0.00	₹ 0.00	₹ 0.00
Salvage	₹ -4,18,423.72	₹ 0.00	₹ -1,78,171.15	₹ 0.00	₹ -2,15,290.64	₹ -24,961.93
Total	₹ 24,15,05,393.47	₹ 10,80,166.87	₹ 18,77,571.56	₹ 23,38,68,988.07	₹ 42,58,400.89	₹ 4,20,266.07
Annual Cost of System						
Capital	₹ 4,02,482.09	₹ 82,722.30	₹ 87,023.68	₹ 0.00	₹ 2,08,555.14	₹ 24,180.98
Replacement	₹ 98,743.84	₹ 0.00	₹ 0.00	₹ 0.00	₹ 88,484.49	₹ 10,259.36
O&M	₹ 1,81,40,718.41	₹ 833.33	₹ 75.00	₹ 1,80,90,790.08	₹ 49,020.00	₹ 0.00
Fuel	₹ 71,922.02	₹ 0.00	₹ 71,922.02	₹ 0.00	₹ 0.00	₹ 0.00
Salvage	₹ -32,366.91	₹ 0.00	₹ -13,782.32	₹ 0.00	₹ -16,653.67	₹ -1,930.91
Total	₹ 1,86,81,499.46	₹ 83,555.64	₹ 1,45,238.38	₹ 1,80,90,790.08	₹ 3,29,405.95	₹ 32,509.42

TABLE 7.23
Net Present and Annual Cost with tariff-2 for Proposed Microgrid without DR Events

Cost Components	System	Generic Flat Plate PV	Diesel Generator	Grid Tariff-2	Battery Storage	System Converter
Net Present Cost of System						
Capital	₹ 55,20,706.44	₹ 10,69,393.94	₹ 11,25,000.00	₹ 0.00	₹ 30,33,250.00	₹ 2,93,062.50
Replacement	₹ 14,17,391.73	₹ 0.00	₹ 0.00	₹ 0.00	₹ 12,93,052.98	₹ 1,24,338.75
O&M	₹ 22,92,85,321.06	₹ 10,772.93	₹ 1,551.30	₹ 22,85,60,044.29	₹ 7,12,952.54	₹ 0.00
Fuel	₹ 13,42,865.81	₹ 0.00	₹ 13,42,865.81	₹ 0.00	₹ 0.00	₹ 0.00
Salvage	₹ -4,36,144.08	₹ 0.00	₹ -1,77,272.81	₹ 0.00	₹ -2,35,469.46	₹ -23,401.81
Total	₹ 23,71,30,140.96	₹ 10,80,166.87	₹ 22,92,144.30	₹ 22,85,60,044.29	₹ 48,03,786.06	₹ 3,93,999.44
Annual Cost of System						
Capital	₹ 4,27,050.81	₹ 82,722.30	₹ 87,023.68	₹ 0.00	₹ 2,34,635.17	₹ 22,669.67
Replacement	₹ 1,09,641.46	₹ 0.00	₹ 0.00	₹ 0.00	₹ 1,00,023.31	₹ 9,618.15
O&M	₹ 1,77,36,223.36	₹ 833.33	₹ 120.00	₹ 1,76,80,120.03	₹ 55,150.00	₹ 0.00
Fuel	₹ 1,03,876.55	₹ 0.00	₹ 1,03,876.55	₹ 0.00	₹ 0.00	₹ 0.00
Salvage	₹ -33,737.65	₹ 0.00	₹ -13,712.83	₹ 0.00	₹ -18,214.59	₹ -1,810.23
Total	₹ 1,83,43,054.53	₹ 83,555.64	₹ 1,77,307.40	₹ 1,76,80,120.03	₹ 3,71,593.88	₹ 30,477.58

TABLE 7.24
Net Present and Annual Cost with tariff-1 for Proposed Microgrid with DR Events

Cost Components	System	Generic Flat Plate PV	Diesel Generator	Grid Tariff-1	Battery Storage	System Converter
Net Present Cost of System						
Capital	₹ 52,03,093.94	₹ 10,69,393.94	₹ 11,25,000.00	₹ 0.00	₹ 26,96,100.00	₹ 3,12,600.00
Replacement	₹ 12,76,512.67	₹ 0.00	₹ 0.00	₹ 0.00	₹ 11,43,884.67	₹ 1,32,628.00
O&M	₹ 23,11,71,216.37	₹ 10,772.93	₹ 969.56	₹ 23,05,25,767.02	₹ 6,33,706.86	₹ 0.00
Fuel	₹ 9,29,773.15	₹ 0.00	₹ 9,29,773.15	₹ 0.00	₹ 0.00	₹ 0.00
Salvage	₹ -4,18,423.72	₹ 0.00	₹ -1,78,171.15	₹ 0.00	₹ -2,15,290.64	₹ -24,961.93
Total	₹ 23,81,62,172.41	₹ 10,80,166.87	₹ 18,77,571.56	₹ 23,05,25,767.02	₹ 42,58,400.89	₹ 4,20,266.07
Annual Cost of System						
Capital	₹ 4,02,482.09	₹ 82,722.30	₹ 87,023.68	₹ 0.00	₹ 2,08,555.14	₹ 24,180.98
Replacement	₹ 98,743.84	₹ 0.00	₹ 0.00	₹ 0.00	₹ 88,484.49	₹ 10,259.36
O&M	₹ 1,78,82,105.62	₹ 833.33	₹ 75.00	₹ 1,78,32,177.29	₹ 49,020.00	₹ 0.00
Fuel	₹ 71,922.02	₹ 0.00	₹ 71,922.02	₹ 0.00	₹ 0.00	₹ 0.00
Salvage	₹ -32,366.91	₹ 0.00	₹ -13,782.32	₹ 0.00	₹ -16,653.67	₹ -1,930.91
Total	₹ 1,84,22,886.68	₹ 83,555.64	₹ 1,45,238.38	₹ 1,78,32,177.29	₹ 3,29,405.95	₹ 32,509.42

TABLE 7.25
Net Present and Annual Cost with tariff-2 for Proposed Microgrid with DR Events

Cost Components	System	Generic flat plate PV	Diesel Generator	Grid Tariff-2	Battery Storage	System Converter
Net Present Cost of System						
Capital	₹ 51,83,556.44	₹ 10,69,393.94	₹ 11,25,000.00	₹ 0.00	₹ 26,96,100.00	₹ 2,93,062.50
Replacement	₹ 13,16,536.18	₹ 0.00	₹ 0.00	₹ 0.00	₹ 11,92,197.43	₹ 1,24,338.75
O&M	₹ 22,61,30,165.00	₹ 10,772.93	₹ 1,163.48	₹ 22,54,84,521.73	₹ 6,33,706.86	₹ 0.00
Fuel	₹ 11,24,488.79	₹ 0.00	₹ 11,24,488.79	₹ 0.00	₹ 0.00	₹ 0.00
Salvage	₹ -3,61,992.60	₹ 0.00	₹ -1,77,871.70	₹ 0.00	₹ -1,60,719.09	₹ -23,401.81
Total	₹ 23,33,92,753.81	₹ 10,80,166.87	₹ 20,72,780.56	₹ 22,54,84,521.73	₹ 43,61,285.20	₹ 3,93,999.44
Annual Cost of System						
Capital	₹ 4,00,970.78	₹ 82,722.30	₹ 87,023.68	₹ 0.00	₹ 2,08,555.14	₹ 22,669.67
Replacement	₹ 1,01,839.84	₹ 0.00	₹ 0.00	₹ 0.00	₹ 92,221.69	₹ 9,618.15
O&M	₹ 1,74,92,158.23	₹ 833.33	₹ 90.00	₹ 1,74,42,214.90	₹ 49,020.00	₹ 0.00
Fuel	₹ 86,984.13	₹ 0.00	₹ 86,984.13	₹ 0.00	₹ 0.00	₹ 0.00
Salvage	₹ -28,001.71	₹ 0.00	₹ -13,759.15	₹ 0.00	₹ -12,432.32	₹ -1,810.23
Total	₹ 1,80,53,951.27	₹ 83,555.64	₹ 1,60,338.65	₹ 1,74,42,214.90	₹ 3,37,364.50	₹ 30,477.58

A comparative analysis of the proposed microgrid system for COE, net present cost, operating cost and savings with respect to DR events including different tariffs is shown in Table 7.26, which indicates that the minimum COE (4.24 ₹/kWh), net present cost (₹ 23,33,92,754) and operating cost (₹ 22,61,30,165.0) are achieved with tariff-2 under DR events. The relative graphical representations of COE, operating and net present cost of the base system for both tariffs with DR events are represented in Figure 7.13. Resilience is considered for long-duration outages and reliability for short outages. These are the ability of the system of a facility to respond to an extended, multi-day utility outage and short-duration outage, for example, 2 hr. The assumptions related to outage for the presented case study are discussed in the previous section. The annual operational parameters of the DG set under normal and outage conditions for the proposed microgrid together with both tariffs and DR events are presented in Tables 7.27, 7.28 and 7.29 to maintain resilience (long-duration outage) and reliability (short-duration outage) within the proposed grid-connected microgrid. The performance of the proposed microgrid during outages for resilience

TABLE 7.26

Comparative Summary of Proposed Microgrid for Different Tariffs and DR Events

Parameters	Without DR Events		With DR Events	
	Tariff-1	Tariff-2	Tariff-1	Tariff-2
COE (₹/kWh)	4.37 ₹/kWh	4.30 ₹/kWh	4.32 ₹/kWh	4.24 ₹/kWh
Net present cost (₹)	₹ 24,15,05,393	₹ 23,71,30,140	₹ 23,81,62,172	₹ 23,33,92,754
Operating cost (₹)	₹ 23,45,14,437.42	₹ 22,92,85,321.06	₹ 23,11,71,216.3	₹ 22,61,30,165.0
Utility Bill savings (₹)	₹ 59,08,791	₹ 63,19,461	₹ 59,76,110	₹ 63,66,072
Total Bill savings (₹)	₹ 7,64,00,000	₹ 8,17,00,000	₹ 7,73,00,000	₹8,23,00,000
Payback (yr)	0.65	0.66	0.64	0.60

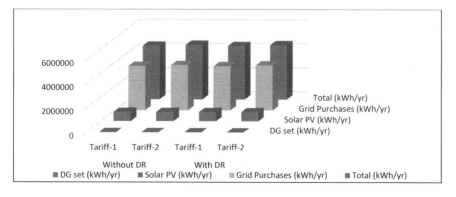

FIGURE 7.13 COE, operating and net present cost of proposed grid-connected microgrid with DR events.

TABLE 7.27
DG Set Operational Parameters for Both Tariffs and DR Events for Proposed Microgrid

Parameters	Without DR Events		With DR Events	
	Tariff-1	Tariff-2	Tariff-1	Tariff-2
Hours of Operation	5.0 hrs/yr	8.0 hrs/yr	5.0 hrs/yr	6.0 hrs/yr
Number of Starts	1.0 starts/yr	3.0 starts/yr	1.0 starts/yr	2.0 starts/yr
Operational Life	3000 yr	1875 yr	3000 yr	2,500 yr
Electrical Production	4,126 kWh/yr	5883 kWh/yr	4,126 kWh/yr	4995 kWh/yr
Mean Electrical Output	825 kW	735 kW	825 kW	833 kW
Fuel Consumption	1,090 L/yr	1,574 L/yr	1,090 L/yr	1,318 L/yr
Generator Fuel Price	66.0 ₹/L	66.0 ₹/L	66.0 ₹/L	66.0 ₹/L

TABLE 7.28
DG Operation for Different Tariffs and DR Events under Outage for Reliability (Short Outage) for Proposed Microgrid

Components	Without DR Events		With DR Events	
	Tariff-1	Tariff-2	Tariff-1	Tariff-2
DG runtime (hours/day)	1.74	2.78	1.82	2.18
DG O&M cost (₹/day)	26.1	41.7	27.3	32.7
DG fuel consumption (L/day)	379	547	396	479
DG fuel cost (₹/day)	25,016	36,131	26,153	31,631
Battery throughput (kWh/day)	1,675	1,549	1,655	1,790

TABLE 7.29
DG Operation for Different Tariffs and DR Events under Outage for Resilience (Long Outage) for Proposed Microgrid

Components	Without DR Events		With DR Events	
	Tariff-1	Tariff-2	Tariff-1	Tariff-2
DG runtime (hours/day)	2.50	2.00	2.50	2.00
DG O&M cost (₹/day)	37.5	30.0	37.5	30.0
DG fuel consumption (L/day)	545	550	545	538
DG fuel cost (₹/day)	35,961	36,276	35,961	35,539
Battery throughput (kWh/day)	1,903	1,627	1,903	1,944

Utility Tariff Variation and Demand Response Events 169

(long outage) and reliability (short outage) without and with DR events are illustrated in Figures 7.14 and 7.15.

The performance of the proposed grid-connected microgrid system during resilience (long outage) and reliability (short outage) is maintained by combined operation of renewable (solar PV) power, battery power and generator power only, which fulfill the load demand under long- and short-duration outages to maintain the resilience and reliability of the system. The pollutant emissions for the proposed microgrid system without DR events such as carbon dioxide and particulate matter are 23,20,887

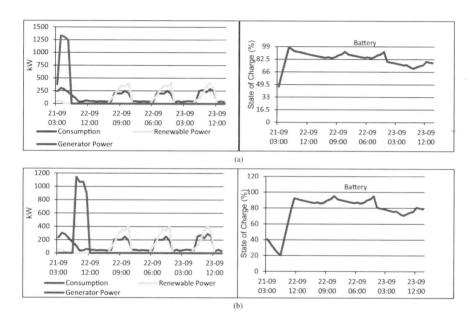

FIGURE 7.14 Proposed microgrid performance during outages for resilience (long outage): (a) without DR events, (b) with DR events.

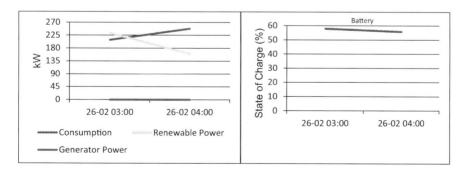

FIGURE 7.15 Proposed microgrid performance during outages for reliability (short outage) with DR events.

kg/yr and 0.109 kg/yr for tariff-1 and 23,70,176 kg/yr and 0.157 kg/yr for tariff-2, and with DR events, these pollutants are 23,14,458 kg/yr and 0.109 kg/yr for tariff-1 and 23,66,200 kg/yr for tariff-2, respectively, which are given in Table 7.30. The details about demand response events and the status of the DG, battery and other parameters under DR events for the proposed microgrid are given in Table 7.31 and Figure 7.16, respectively, which give information on DR event duration and timing and the operational performances of microgrid system components, as well as the load demand reduction under DR events.

The monthly electricity charges/bill breakdown for the proposed grid-connected microgrid without and with DR events are shown in Figure 7.17. The annual savings

TABLE 7.30
Pollutant Emissions of Proposed Microgrid for Both Tariffs with DR Events

Pollutant Emission	Without DR Events		With DR Events	
	Tariff-1	Tariff-2	Tariff-1	Tariff-2
Carbon Dioxide (kg/yr)	23,20,887	23,70,176	23,14,458	23,66,200
Carbon Monoxide (kg/yr)	18.0	26.0	18.0	21.7
Unburned Hydrocarbons (kg/yr)	0.785	1.13	0.785	0.949
Particulate Matter (kg/yr)	0.109	0.157	0.109	0.132
Sulfur Dioxide (kg/yr)	10,057	10,268	10,029	10,252
Nitrogen Oxides (kg/yr)	4,932	5,041	4,918	5,030

TABLE 7.31
Demand Response Events with DG and Battery Status for Proposed Microgrid

Parameters	Event-1 Start: 15 Mar 03:00 p.m.	Event-2 Start: 17 Apr 02:00 p.m.	Event-3 Start: 16 Aug 02:00 p.m.	Event-4 Start: 08 Oct 03:00 p.m.	Event-5 Start: 27 Nov 02:00 p.m.	Event-6 Start: 24 Dec 01:00 p.m.
	Value	Value	Value	Value	Value	Value
Duration (hrs)	3	3	3	3	3	3
Optimized demand reduction bid (kW)	1120	1180	1180	922	996	911
Demand reduction revenue (₹)	39,239	41314	41195	32274	34869	31901
Diesel generator operating hours	0	0	0	0	0	0
Battery throughput (kWh)	287	242	390	322	329	165
Battery wear cost (₹)	55.4	46.7	75.3	62.1	63.6	31.8

Utility Tariff Variation and Demand Response Events

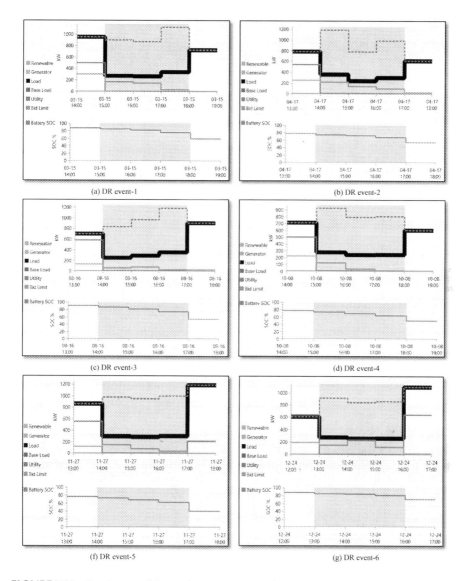

FIGURE 7.16 Implemented demand response events for proposed grid-connected microgrid.

cost summary and different cast flows during the lifetime of the proposed microgrid system, including DR events (with and without DR), are illustrated in Figures 7.18 and 7.19. In Figure 7.18, the chart summarizes the estimated annual savings in the following categories:

- Demand: Savings from demand charge reduction
- Energy: Consumption reduction and self-consumption
- Outages: Cost savings related to utility service interruptions

172 Microgrids

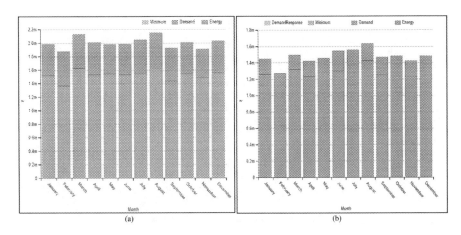

FIGURE 7.17 Monthly electricity bill breakdown for proposed grid-connected microgrid: (a) without DR events, (b) with DR events.

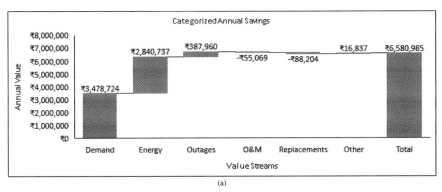

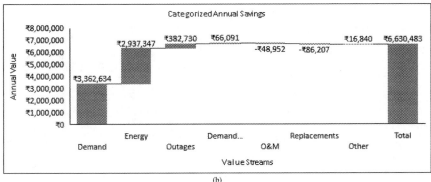

FIGURE 7.18 Annual savings cost summary for proposed grid-connected microgrid: (a) without DR events, (b) with DR events.

Utility Tariff Variation and Demand Response Events 173

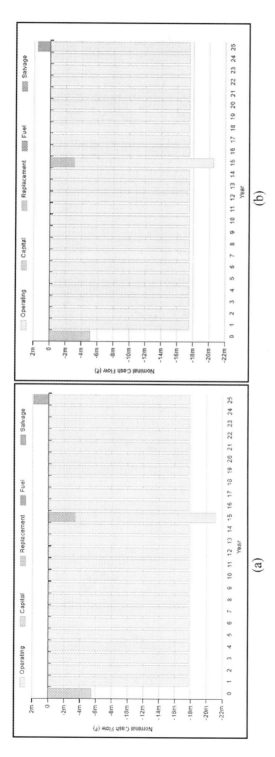

FIGURE 7.19 Different cash flow during the lifetime of the proposed microgrid system: (a) without DR events, (b) with DR events.

- O&M: Operating and maintenance costs of the proposed components
- Replacements: Cost to replace proposed system components over the project lifetime
- Other: Cost differences not included in the other categories
- Total: The total savings (annualized) of the proposed system

The different system component output power profiles and utility bill comparison with the base system and proposed microgrid system without and with DR events are illustrated in Figures 7.20 and 7.21, which clearly show the detailed performance of different system component output power profiles throughout the year. The monthly peak day profiles of proposed microgrid components, including DR events (with and without DR events), are demonstrated in Figures 7.22 and 7.23, which show the overall performance of the proposed grid-connected microgrid system with the effect of DR events in different operating conditions.

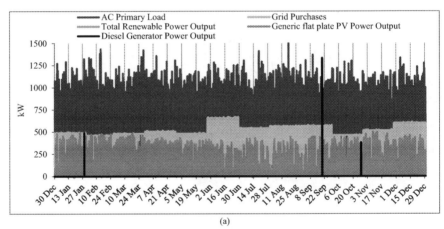

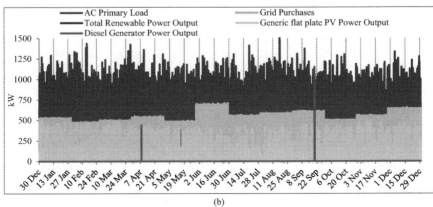

FIGURE 7.20 Different system component output power profiles for the proposed grid-connected microgrid: (a) without DR events, (b) with DR events.

Utility Tariff Variation and Demand Response Events 175

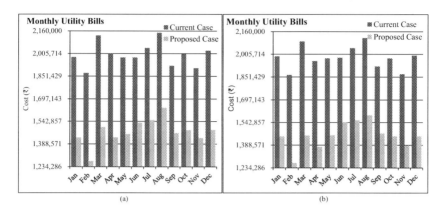

FIGURE 7.21 Monthly peak day profiles of proposed microgrid components with DR events.

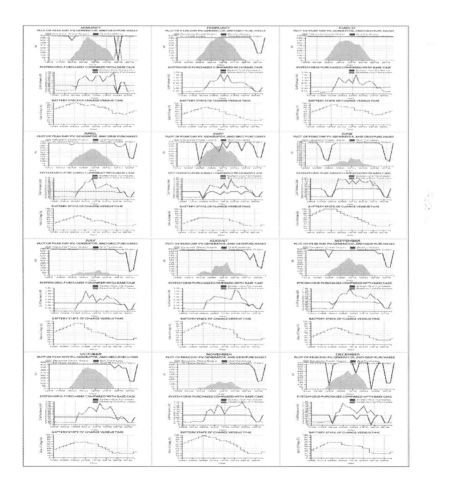

FIGURE 7.22 Monthly peak day profiles of proposed microgrid components without DR events.

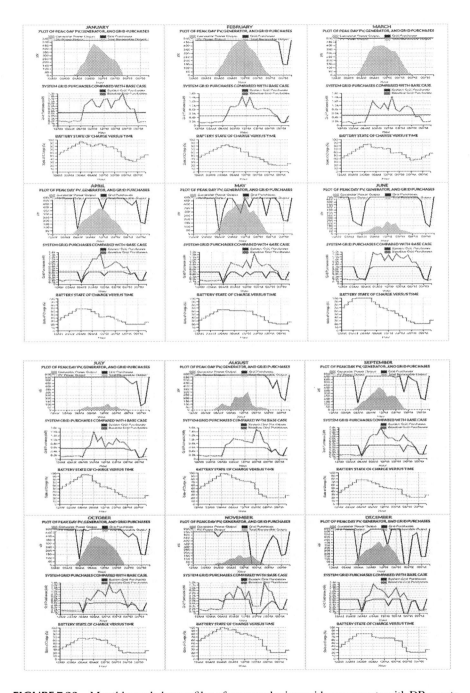

FIGURE 7.23 Monthly peak day profiles of proposed microgrid components with DR events.

The comparative summary between the base case system and best feasible solution for the proposed grid-connected microgrid system for the NITD system can be presented on the basis of the minimum net present cost and cost of energy, which clearly indicates that the net present cost of the proposed microgrid connected to the grid system is significantly reduced to ₹ 23,33,92,754 with DR and ₹ 23,71,30,140 without DR, while it was ₹ 31,78,10,200 for the base case system. Also, the cost of energy of the system decreases noticeably, from ₹ 5.76/kWh (for the base system) to 4.24 ₹/kWh with DR and 4.30 ₹/kWh without DR for the proposed microgrid system, respectively. Similarly, the total annual utility bill savings and demand charge savings for the proposed grid-connected microgrid system are significantly reduced as compared to the base system. Pollutant emissions like CO_2 and particulate matter for the proposed microgrid system are considerably decreased, to 23,70,176 kg/yr and 0.157 kg/yr without DR and 23,66,200 kg/yr and 0.132 kg/yr with DR, respectively, compared to base case system. The comparison of utility bills and the cash flow of net present cost during the project lifetime for the current/base system and the proposed grid-connected microgrid system confirms that the proposed microgrid system has significant impacts on base system performance considering DR events for different tariff schemes of utilities. Finally, overall performance is enhanced with the proposed grid-connected microgrid system.

7.6 CONCLUSIONS

The proposed grid-connected microgrid system design consisting of solar PV, battery storage and diesel generator connected with the DVC-NITD-grid is analyzed based on various system performances. The resulting analysis of the proposed case study is divided into two cases, the base case system and the proposed grid-connected microgrid system, which are scrutinized using HOMER-grid for two options of tariff by the utility considering demand response events. The base case system or current system of NITD presently spends approximately ₹ 2,39,99,581.82 on the utility bill annually in which ₹ 57,23,737.24 is demand charges, approximately 24% of the utility bill. Also, the total energy purchasing charges are ₹ 1,82,75,844.59, which is a major share of the total utility bill. The proposed grid-connected microgrid system reduces the system annual utility bill to ₹ 1,76,80,120.03 without DR events and ₹ 1,74,42,214.90 with DR events by considering tariff-2 as compared to the base case system. Correspondingly, the demand and energy purchase charges are also reduced with the proposed microgrid system compared to the base system. The superior design of the proposed grid-connected microgrid is obtained with tariff-2 for both conditions of DR events. As a result of that, the net present cost of the proposed grid-connected microgrid system is significantly reduced to ₹ 23,33,92,754 with DR and ₹ 23,71,30,140 without DR as compared to the base case system (i.e., ₹ 31,78,10,200). Also, the cost of energy of the system is decreased noticeably from ₹ 5.76/kWh (for base system) to 4.24 ₹/kWh with DR and 4.30 ₹/kWh without DR for the proposed microgrid system.

Similarly, the total annual utility bill savings and demand charge savings for the proposed grid-connected microgrid system are significantly reduced compared to the base system. Pollutant emissions like CO_2 and particulate matter for the proposed microgrid system are considerably decreased to 23,70,176 kg/yr and 0.157 kg/yr without DR and 23,66,200 kg/yr and 0.132 kg/yr with DR, respectively, from the base case system. The comparative analysis of utility bills and the net present cost cash flow during the project lifetime for the current system and proposed grid-connected microgrid system confirms that the proposed microgrid system has noteworthy positive influences on system performances. The reliability and resiliency of the proposed grid-connected microgrid system are enhanced compared to the base system while considering short outages for reliability and long outages for resilience for both tariffs and demand response. Also, in the proposed grid-connected microgrid system, there are more options to supply the load demand efficiently; hence, system reliability and resilience are enhanced. The required DG operation hours and net grid energy purchases are reduced; therefore, pollutant emission components such as CO_2 and particulate matter are decreased for the proposed microgrid system. The lower pollutant emissions decrease the environmental and social impacts and provides for better human life regarding the bad impacts of pollutants on health. Finally, it is concluded that the proposed design of the grid-connected microgrid system is more feasible with respect to the base or current system of the case study for both conditions of DR events with tariff-2, and the overall performance of the proposed system is improved compared to the base system.

7.7 ACKNOWLEDGMENTS

First, the authors would like to give thanks to NITD for helping and providing the facility to conduct research on the NITD system. Moreover, the authors also would like to acknowledge HOMER Energy for the HOMER-grid software and the NASA Prediction of Worldwide Energy Resource (Power) database to obtain the solar data of steady locations.

REFERENCES

1. Paterakis, N.G., Erdinç, O. and Catalão, J.P., 2017. An overview of demand response: Key-elements and international experience. *Renewable and Sustainable Energy Reviews*, 69, pp. 871–891.
2. Chai, Y., Xiang, Y., Liu, J., Gu, C., Zhang, W. and Xu, W., 2019. Incentive-based demand response model for maximizing benefits of electricity retailers. *Journal of Modern Power Systems and Clean Energy*, 7(6), pp. 1644–1650.
3. He, L., Zhang, S., Chen, Y., Ren, L. and Li, J., 2018. Techno-economic potential of a renewable energy-based microgrid system for a sustainable large-scale residential community in Beijing, China. *Renewable and Sustainable Energy Reviews,* 93, pp. 631–641.
4. Diou, C., Chrysopoulos, A., Symeonidis, A.L. and Mitkas, P.A., 2016. Response modeling of small-scale energy consumers for effective demand response applications. *Electric Power Systems Research*, 132, pp. 78–93.
5. Aghaei, J., Alizadeh, M.I., Siano, P. and Heidari, A., 2016. Contribution of emergency demand response programs in power system reliability. *Energy*, 103, pp. 688–696.

6. Tang, C.J., Dai, M.R., Chuang, C.C., Chiu, Y.S. and Lin, W.S., 2014. A load control method for small data centers participating in demand response programs. *Future Generation Computer Systems*, 32, pp. 232–245.
7. Babar, M., Ahamed, T.I., Al-Ammar, E.A. and Shah, A., 2013. A novel algorithm for demand reduction bid based incentive program in direct load control. *Energy Procedia*, 42, pp. 607–613.
8. Masrur, H., Howlader, H.O.R., Elsayed Lotfy, M., Khan, K.R., Guerrero, J.M. and Senjyu, T., 2020. Analysis of techno-economic-environmental suitability of an isolated microgrid system located in a remote Island of Bangladesh. *Sustainability*, 12(7), p. 2880.
9. Safamehr, H. and Rahimi-Kian, A., 2015. A cost-efficient and reliable energy management of a micro-grid using intelligent demand-response program. *Energy*, 91, pp. 283–293.
10. Suresh, V., Muralidhar, M. and Kiranmayi, R., 2020. Modelling and optimization of an off-grid hybrid renewable energy system for electrification in a rural areas. *Energy Reports*, 6, pp. 594–604.
11. Mazzola, S., Astolfi, M. and Macchi, E., 2016. The potential role of solid biomass for rural electrification: A techno economic analysis for a hybrid microgrid in India. *Applied Energy*, 169, pp. 370–383.
12. Tudu, B., Mandal, K.K. and Chakraborty, N., 2019. Optimal design and development of PV-wind-battery based nano-grid system: A field-on-laboratory demonstration. *Frontiers in Energy*, 13(2), pp. 269–283.
13. Zhang, D., Evangelisti, S., Lettieri, P. and Papageorgiou, L.G., 2015. Optimal design of CHP-based microgrids: Multiobjective optimisation and life cycle assessment. *Energy*, 85, pp. 181–193.
14. Kumar, P. and Palwalia, D.K., 2018. Feasibility study of standalone hybrid wind-PV-battery microgrid operation. *Technology and Economics of Smart Grids and Sustainable Energy*, 3(17), pp. 1–16.
15. García, P., Torreglosa, J.P., Fernández, L.M., Jurado, F., Langella, R. and Testa, A., 2016. Energy management system based on techno-economic optimization for microgrids. *Electric Power Systems Research*, 131, pp. 49–59.
16. Thomas, D., Deblecker, O. and Ioakimidis, C.S., 2016. Optimal design and techno-economic analysis of an autonomous small isolated microgrid aiming at high RES penetration. *Energy*, 116, pp. 364–379.
17. Sawle, Y., Gupta, S.C. and Bohre, A.K., 2018. Review of hybrid renewable energy systems with comparative analysis of off-grid hybrid system. *Elsevier, Renewable and Sustainable Energy Reviews*, 81, pp. 2217–2235, 2017.
18. Guo, L., Wang, N., Lu, H., Li, X. and Wang, C., 2016. Multi-objective optimal planning of the stand-alone microgrid system based on different benefit subjects. *Energy*, 116, pp. 353–363.
19. Elavarasan, R.M., 2019. The motivation for renewable energy and its comparison with other energy sources: A review. *European Journal of Sustainable Development Research*, 3(1), pem0076(1–19).
20. Babatunde, O.M., Munda, J.L. and Hamam, Y., 2020. A comprehensive state-of-the-art survey on hybrid renewable energy system operations and planning. *IEEE Access*, 8, pp. 75313–75346.
21. HOMER energy. HOMER grid v1.8 user manual, 2020. www.homerenergy.com/products/grid/docs/1.8/design.html.
22. Go, R., Kahrl, F. and Kolster, C., 2020. Planning for low-cost renewable energy. *The Electricity Journal*, 33(2), p. 106698(1–5).

23. Fan, X., Sun, H., Yuan, Z., Li, Z., Shi, R. and Razmjooy, N., 2020. Multi-objective optimization for the proper selection of the best heat pump technology in a fuel cell-heat pump micro-CHP system. *Energy Reports*, 6, pp. 325–335.
24. Bohre, A.K., Agnihotri, G. and Dubey, M., 2016. Optimal sizing and sitting of DG with load models using soft computing techniques in practical distribution system. *IET Generation, Transmission & Distribution*, 10(6), pp. 1–16.
25. Dali, M., Belhadj, J. and Roboam, X., 2010. Hybrid solar–wind system with battery storage operating in grid-connected and standalone mode: Control and energy management—experimental investigation. *Energy*, 35(6), pp. 2587–2595.
26. Saib, S. and Gherbi, A., 2015, May. Simulation and control of hybrid renewable energy system connected to the grid. In 2015 5th International Youth Conference on Energy (IYCE), IEEE, pp. 1–6.
27. Khalid, M., AlMuhaini, M., Aguilera, R.P. and Savkin, A.V., 2018. Method for planning a wind–solar–battery hybrid power plant with optimal generation-demand matching. IET. *Renewable Power Generation*, 12(15), pp. 1800–1806.
28. DVC: Tariff order of DVC for the years 2015–16 and 2017–18. http://wberc.gov.in/sites/default/files/Tariff%20Order_DVC_2014-2017_pdf%20%281%29.pdf.

8 SOS-Based Load Frequency Controller Design for Microgrids Using Degrees-of-Freedom PID Controller

Sunita Pahadasingh, Chitralekha Jena, and Chinmoy Kumar Panigrahi

CONTENTS

8.1 Introduction 181
8.2 Proposed Model 183
    8.2.1 Three-Area System 183
    8.2.2 Controller Design... 186
        8.2.2.1 PID Controller.. 186
        8.2.2.2 Two-Degrees-of-Freedom Controller 187
        8.2.2.3 Three-Degrees-of-Freedom Controller............................ 188
    8.2.3 Symbiotic Organism Search Algorithm 189
8.3 Simulation Results 189
8.4 Conclusion 197
8.5 Appendix 198
References.. 198

8.1 INTRODUCTION

The control to achieve zero steady-state errors for tie lines and keep system frequency to its nominal value is the major challenge in an interconnected power system. During continuous load fluctuations, primary control by a seed governor balances active power generation but is not suitable for controlling frequency. Hence, to keep system frequency closer to minimal values, an improved control technique called load frequency control is analyzed. The concept of LFC was originated by Concordia [1]. In

DOI: 10.1201/9781003121626-8

the past, a large amount of research was done for single-area and interconnected LFC systems. Most studies discussed thermal unit generation. A single-area LFC system was proposed by Wang et al. [2]. A two-area interconnected thermal system was developed by Elgard and Fossa [3]. A three-area unequal thermal plant was discussed by Debbarma et al. [4], and Rahman et al. explored four-area unequal thermal plants [5]. In [6], interconnected two-area thermal hydro-plants were elaborated upon by Kothari and Nanda. Monanty et al. presented two-area interconnected thermal hydro and gas power plants [7]. A few of these are related to LFC of an interconnected multi-area power system considering non-linearity [8] in terms of generation rate constraint (GRC) and governor dead band (GDB). GRC is a physical restraint to keep the degree of transformation of turbine power within a specified limit, generally 3% per minute, described in [9–10]. In [11], a GDB of 0.06% is reflected for the scheme and exploration drive of an LFC system.

In the past decade, power generation has mostly depended on thermal and hydro power plants. With the increase in population, industries demand more power, but the availability of coal is going to be exhausted within a very short duration. As fossil fuels are exhausted with growing demand, conventional units are being integrated with distributed energy resources located closer to load centers, termed microgrids. Currently, a hybrid interconnected power system comprising conventional units and renewable energy sources like solar, wind and so on is an extensive research area in LFC [12–17]. This system diminishes transmission losses and delivers improved controllability by filling power demands. However, with irregular types of renewable sources, along with microgrids, system frequency divergences are significantly high. A microgrid could be designed in the form of distributed generation sources such as a wind turbine generator, fuel cell, aqua electrolyzer, diesel engine generator (DEG) and battery energy storage system. To sustain proper steadiness between real power and demand, a suitable control strategy is needed for power generation from these microgrid resources. Microgrid frequency deviates due to the interconnection of these energy resources. Since wind power generation is intermittent in nature, a fraction of WTG power is used for hydrogen production in AE [14–15]. The system behavior of a WTG differs from conventional thermal units due to the intermittent power curve. Hence WTGs and BESSs are harmonized to regulate the frequency eccentricities proposed by Abdul et al. [18]. For long durations, fuel cells and fly wheels can condense frequency fluctuations, whereas energy storage devices are considered for short duration, discussed by Vidyanandan et al. [19]. BESSs [20] with multiple sources are used to soothe system frequency deviation by leveling the load with respect to generation. In [21], BESS optimization is generally used as primary reserve for minimum frequency deviation in microgrid sources. BESS discharges if the system frequency is higher than the scheduled value and charges at a state of lower value as compared to the nominal value.

Many researchers have already discussed conventional controllers [22] such as proportional integral (PI) and proportional integral derivative controllers and artificial neural fuzzy intelligence systems (ANFISs) [23]. Mostly, the PID controller is used in LFC for simplification, and fewer parameters are to be optimized [24–25]. In papers [26–28], a fuzzy logic PID control for LFC has better robustness as compared to conventional PIDs. However, such methods suffer limitations of large computational time due to the design process of membership function. In industry applications, the PID controller is mostly considered for its simpler design and robust nature. However, during transient periods, it suffers from high settling time and large overshoot. Through multiple control loops, control action is achieved, which is basically degrees of freedom [29]. In [30–33], the concept of a 2DOF-PID controller was discussed that has improved performance characteristics likened to conventional controllers. Raju et al. [31] proposed 2DOF-PID control strategies for an LFC system. Hence, when there are more tuning knobs present in a controller, the performance of the latter is better in LFC. A three degree-of-freedom PID controller can also be experimented with for operative use in the proposed system [34–35]. Rahman et al. [35] describe the supremacy of the 3DOF-ID controller over 2DOF-ID and single-degree-of-freedom PIDs using the biogeography-based optimization (BBO) technique. The effect of a 3DOF-PID with a dish-Stirling solar system was discussed in [12] by Rahman et al. The addition of a DC link with an existing AC link has better stability, as discussed by Ibraheem et al. [36]. A high-voltage direct current link has less frequency oscillation and better transient stability and settles down quickly to achieve a steady state, as discussed in [36–39].

Gain parameters of secondary controllers are generally adjusted by optimization techniques. In this area, many research has implemented different optimization techniques such as biogeography-based optimization [5, 12 and 35], differential evolution (DE) [7, 11], flower pollination (FP) [13], particle swarm optimization (PSO) [14], hybrid DE-PSO [27, 28], grey wolf optimization (GWO) [24, 29], bacteria foraging optimization algorithm (BFOA) [25], cuckoo search (CS) [33], teaching learning-based optimization [26, 34, 37 and 40], salp swarm algorithm (SSA) [41], ant lion optimizer (ALO) [42] and symbiotic organism search [31, 43, 44]. For optimization of controller parameters, an objective function is necessary, which is basically in the time domain. Different performance indexes such as integral of absolute error, integral of square error, integral of time multiplied square error and integral of time multiplied absolute error (ITAE) are considered as objective functions to minimize the area control error. ITAE is mostly considered as objective function because its transient behavior is superior compared to others.

The main contributions of this chapter are as follows.

- Design of three area hybrid thermal and hydro power sources with microgrid energy resources applied to area1.
- Secondary controllers PID, 2DOF-PID and 3DOF-PID are considered to minimize area control error, and the dynamic performances of these controllers are compared through numerous simulations.
- Controller parameters are optimized using SOS, TLBO and DEPSO algorithms, and the results are compared simultaneously.

8.2 PROPOSED MODEL

8.2.1 Three-Area System

In this chapter, a three-area LFC system model is used, which consists of thermal units and hydro units as multiple sources in area 1, area 2 and area 3. Here, in each control area, system non-linearities are taken, such as GDB, GRC and reheat turbine

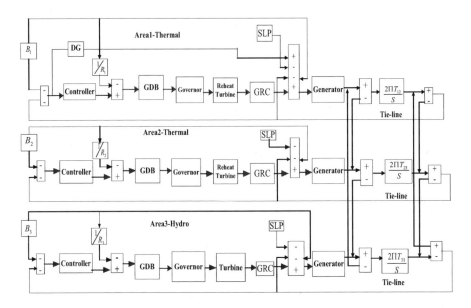

FIGURE 8.1 Representation of Simulink model of power system under study.

connected by means of AC tielines. Distributed generation resources are applied to area 1, which consists of a WTG, FC, AE, DEG and BESS.

Area control error (ACE) is taken as input to each controller, and the general expression of this is given in Eqs. (8.1), (8.2) and (8.3) [45].

$$ACE_1 = B_1 \Delta F_1 + \Delta P_{12} \tag{8.1}$$

$$ACE_2 = B_2 \Delta F_2 + \Delta P_{23} \tag{8.2}$$

$$ACE_3 = B_3 \Delta F_3 + \Delta P_{31} \tag{8.3}$$

An impartial generating unit sited downstream of the electric distribution system at or near the end user is known as distributed generation resources of microgrid. Not only does distributed generation meet a specific customer's energy needs, it can also support the economic operation of the transmission and distribution grid. Currently, distributed generation is used for (1) standby or emergency power, (2) cogeneration (combined heat and power), (3) peak shaving, (4) grid support or (5) standalone on-site power. In this chapter, the following distributed generation resources are discussed briefly and a model of the DGR system is shown in Figure 8.2 [42].

The overall power provided to the load from the hybrid microgrid system is expressed in Eq. (8.4).

$$P_s = P_{WTG} + P_{FC} + P_{AE} - P_{DEG} \pm P_{BESS} \tag{8.4}$$

where P_s is total power supplied, P_{WTG} is the output power of the wind turbine generator, P_{FC} is the output power of the fuel cell, P_{AE} is the output power of the aqua

SOS-Based LFC Design for Microgrids

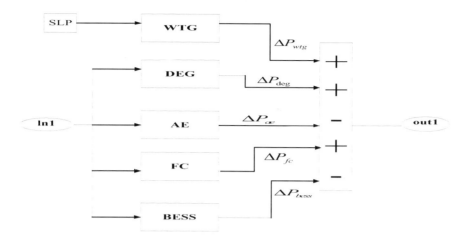

FIGURE 8.2 Model of DGR system.

electrolyzer, P_{DEG} is the output power of the diesel generator and P_{BESS} is the output power of the battery energy storage system.

1. *Diesel engine generator*: DEG consists of an IC engine to adapt fuel energy into mechanical energy and then convert mechanical energy into electrical form through the generator. Hence, the main function of the DEG is to provide good reimbursement after power failure due to technical difficulties. The transfer function of the DEG is given as

$$C_{DEG}(s) = \frac{K_{DEG}}{1 + T_{DEG}S} \tag{8.5}$$

2. *Fuel cell*: This device provides electric energy after conversion from chemical to electrical form through hydrogen storage. It is expected that fuel cells have higher efficiencies and lower capital cost than existing units with less emission. The transfer function of FC is given as

$$C_{FC}(s) = \frac{K_{FC}}{1 + T_{FC}S} \tag{8.6}$$

3. *Wind turbine generator*: This is the fastest-growing renewable energy source due to economically feasibility. The amount of kinetic energy generated depends on the air density and wind speed. Wind power is mainly dependent on wind speed (m^3/s), which varies constantly with reference to time. The transfer function of the wind generator is given as

$$C_{WTG}(s) = \frac{K_{WTG}}{1 + T_{WTG}S} \tag{8.7}$$

4. *Aqua electrolyzer*: Due to high sporadic power generation from wind and photovoltaics in the microgrid system, power vacillation increases. When the fuel cell is equipped with an aqua electrolyzer (AE), the power absorbed by the AE can be controlled to lessen power fluctuations. The transfer function of the aqua electrolyzer is

$$C_{AE}(s) = K_{AE} / 1 + T_{AE}S \tag{8.8}$$

5. *Battery energy storage system*: The basic idea is to convert electricity into an electrochemical form and save it as electrolytes inside a cell. While discharging, the electrolytes react with the electrodes in the cell, and the reverse reaction generates electric current. It provides additional damping oscillations to improve the stability of the power system. The first access and higher energy density characteristics make this more effective in a storage system. The transfer function of the battery energy storage system is

$$C_{BESS}(s) = K_{BESS} / 1 + T_{BESS}S \tag{8.9}$$

8.2.2 Controller Design

8.2.2.1 PID Controller

It is mostly used in an interconnected power system due to simpler structure to minimize the area control error. The basic structure of a PID controller is shown in Figure 8.3 [24]. In this, there are three parameters to be optimized (K_P, K_I, K_D). However, it has severe oscillations, with ultimate overrun and large settling times damaging the system performance.

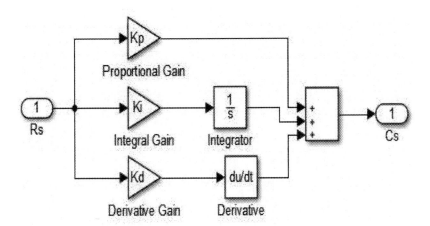

FIGURE 8.3 Basic structure of PID controller.

SOS-Based LFC Design for Microgrids

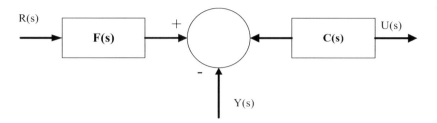

FIGURE 8.4 Control structure of 2DOF system.

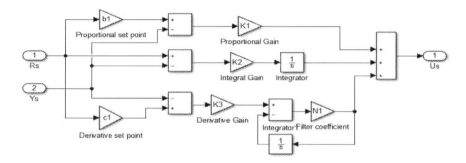

FIGURE 8.5 Basic structure of 2DOF-PID controller.

8.2.2.2 Two-Degrees-of-Freedom Controller

Control action can be achieved by multiple control loops, which are called degrees of freedom. A 2DOF controller, shown in Figure 8.4 [31], is implemented in this system, which is preferable to a PID controller in terms of dynamic response. A 2DOF-PID controller contains two control loops which are optimized for this system. Therefore, when there are more tuning knobs in a controller, the performance of the latter is better in LFC. The main application of a 2DOF controller in control system engineering is to enhance control quality by means of good set point stalking and improved disruption rejection. For a 2DOF-PID controller, $F(s)$ and $C(s)$ are given by Eqs. (8.10) and (8.11).

$$F(s) = \frac{(bK_{Pi} + CK_{Di}N_i)s^2 + (bK_{Pi}N_i + K_{Ii})s + K_{Ii}N_i}{(K_{Pi} + K_{Di}N_i)s^2 + (K_{Pi}N_i + K_{Ii})s + K_{Ii}N_i} \quad (8.10)$$

$$C(s) = \frac{(bK_{Pi} + CK_{Di}N_i)s^2 + (bK_{Pi}N_i + K_{Ii})s + K_{Ii}N_i}{s(s + N_i)} \quad (8.11)$$

where $R(s)$ is the reference signal; $Y(s)$ is the measured output signal; $U(s)$ is the output signal; $C(s)$ is the 1DOF controller; $F(s)$ is the pre-filter on $R(s)$; b is the proportional set-point weighting; c is the derivative set point weighting; and K_p, K_i and K_d are the proportional, integral and derivative gain, respectively.

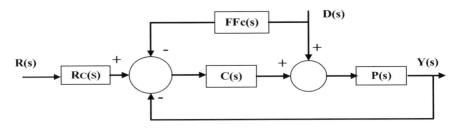

FIGURE 8.6 Control structure of 3DOF system.

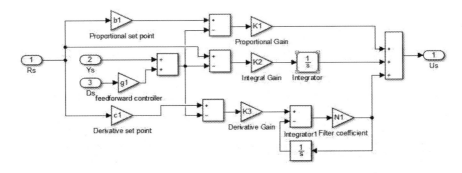

FIGURE 8.7 Structure of 3DOF-PID controller.

8.2.2.3 Three-Degrees-of-Freedom Controller

A three-degrees-of-freedom (3DOF) controller contains three control loops: closed loop stability, modeling of responses and eradication of instabilities, where $R(s)$ is the is the input fed by ACE signal, $Y(s)$ is the measured output signal, $D(s)$ is the system disturbance, $P(s)$ is the plant model, $C(s)$ is the 1DOF controller, $R_C(s)$ is the input reference controller and $FF_C(s)$ is the feed-forward controller.

Closed loop stability is provided by $C(s)$, and the disturbance factor $D(s)$ can be eliminated by the feed-forward controller $FF_C(s)$. The structure of a 3DOF-PID controller is presented in Figure 8.7 [35].

$$Y(s) = \left[\frac{C(s)P(s)R_c(s)}{1+c(s)P(s)}\right]R(s) + \left[\frac{P(s)-C(s)P(s)FF_c(s)}{1+C(s)P(s)}\right]D(s) \quad (8.12)$$

K_{pi}, K_{Ii} and K_{Di} are the gain of the single-order degree of freedom from $C(s)$. N_i is the filter constant for derivative gain. $R_C(s)$ consists of proportional set point weighting b_i and derivative set point weighting. The feed forward controller $FF_C(s)$ has gain parameter g_i.

8.2.3 SYMBIOTIC ORGANISM SEARCH ALGORITHM

This algorithm is the current vigorous meta-heuristic algorithm offered by Cheng and Prayogo [43]. In this technique, no specific algorithm parameters are needed; it only mimics the symbiotic relationships among different organisms for persisting in the ecosystem. In SOS, based on population space, an ecosystem is considered and, through biological interaction among organisms, new populations can be generated. There are three phases in the SOS algorithm: mutualism, commensalism and parasitism. The flowchart of an SOS algorithm is given in Figure 8.8 [31].

1. Mutualism phase: This phase reveals the mutual benefit symbiotic relationship between two different species. In this phase, X_i and X_j are the two arbitrary organisms. To enhance the chance of survival, they interact between themselves

$$\text{(ii)} \quad X_{i,new} = X_i + rand(0,1)*(X_{best} - Mutual\_vector * BF_1) \tag{8.13}$$

$$X_{j,new} = X_j + rand(0,1)*(X_{best} - Mutual_{vector} * BF_2) \tag{8.14}$$

where $Mutual\_vector = \dfrac{X_i + X_j}{2}$ \hfill (8.15)

rand (0,1) is the random number and BF_1, BF_2 are the benefit factor within the range of 1 to 2.

2. Commensalism phase: Two random organisms, X_i and X_j from the ecosystem, are permitted to interrelate in this phase. In this communication organism, X_i assists the interaction, but organism X_j neither assists nor rejects the connection. The new updated value of X_i is calculated

$$X_{i,new} = X_i + rand(-1,1)*(X_{best} - X_j) \tag{8.16}$$

3. Parasitism phase: In this phase, one species gets benefits from the ecosystem, and the other is actively harmed. X_j is selected as a host for the parasite vector from the ecosystem. This vector tries to replace X_j for survival in the ecosystem, and the fitness values of both are calculated.

If the parasite vector has a better fitness value, it will kill X_j from the ecosystem and take its place. If X_j is better, it gets immunity from the parasite vector. Now the parasite vector will no longer be alive in the ecosystem.

8.3 SIMULATION RESULTS

The three-area hybrid source LFC model is designed in MATLAB/Simulink. The microgrid is considered only in area 1. The system performance is evaluated with 1% SLP in area 1 and also for a wind turbine generator. First, the supplementary PID, 2DOF-PID and 3DOF-PID controllers are taken independently. The constraints of the PID controller are first optimized by the DEPSO, TLBO and SOS algorithms.

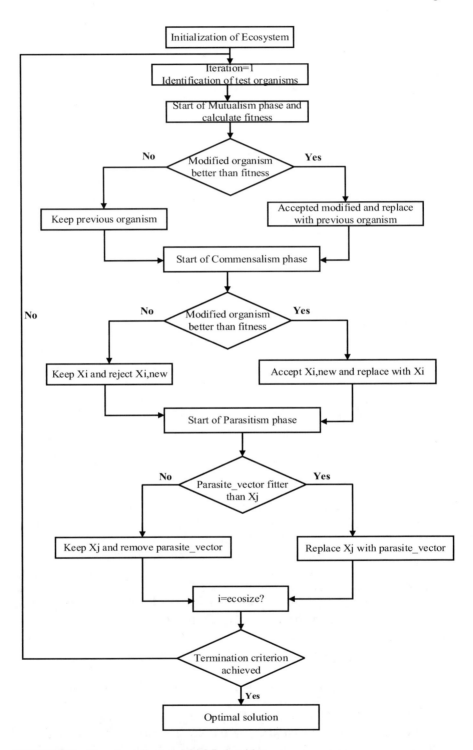

FIGURE 8.8 Flowchart diagram of SOS algorithm.

SOS-Based LFC Design for Microgrids

The corresponding dynamic behaviors are designed and compared in Figures 8.9 and 8.10.

From the tabulated data Table 8.1 and also simulation results, the dynamic behaviors of SOS based PID controller has less undershoot and overshoot as compared to other algorithms such as TLBO and DEPSO.

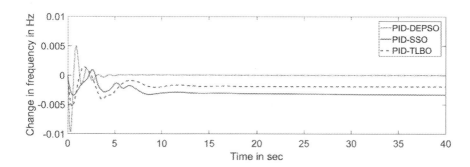

FIGURE 8.9 Frequency deviation using PID controller.

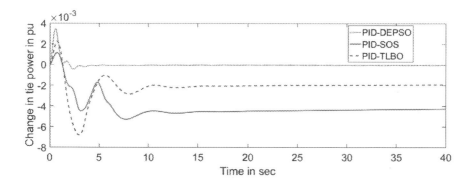

FIGURE 8.10 Tie-line power deviation using PID controller.

TABLE 8.1
Performance Values of PID Controller Using Different Algorithms

Algorithms	Performances	Δf_1 in hz	Δf_2 in hz	Δf_3 in hz	Δf_4 in hz	Δf_5 in hz	Δf_6 in hz
SOSPID	Minimum	−5.338	−4.5666	−6.7907	−3.9039	−0.4790	−6.2950
	Maximum	1.4127	0.4476	2.3093	3.5159	5.5446	2.0698
TLBO-PID	Minimum	−6.4816	−4.6527	−7.0166	−4.2944	−1.3595	−6.8070
	Maximum	2.2035	0.4740	3.7907	4.7906	5.5979	2.9469
DEPSO-PID	Minimum	−9.7230	−5.3470	−7.6769	−4.4739	−1.9072	−7.4412
	Maximum	4.8107	1.3059	4.4167	5.1112	5.9469	3.4517

A comparison of the performance of different controllers, such as PID, 2DOF-PID and 3DOF-PID, is done individually through numerous simulations. The gain parameters of these controllers are optimized by the SOS algorithm. The PID controller has three gain parameters, K_p, K_i and K_d, which is simpler in nature. The 2DOF controller has five gain parameters, K_p, K_i, K_d, b_1 and c_1. This controller has two extra tuning parameters, which improves the transient behavior of the system. The 3DOF-PID controller has six controller parameters, K_p, K_i, K_d, b_1, c_1, and the disturbance factor (d_s). This controller has more tuning parameters as compared to the PID and 2DOF-PID controllers; hence the stability of dynamic performances are superior compared to others shown in the simulation results (Figures 8.11–8.16). The performance values of the controllers are depicted in Table 8.2.

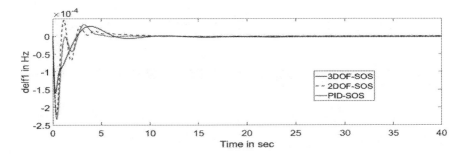

FIGURE 8.11 Frequency deviation for area 1.

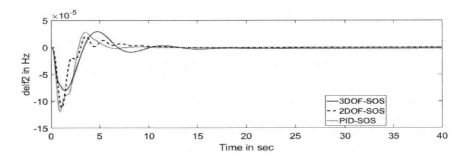

FIGURE 8.12 Frequency deviation for area 2.

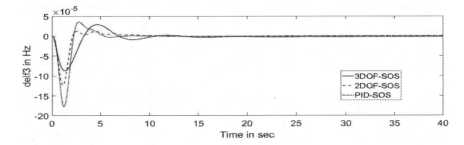

FIGURE 8.13 Frequency deviation for area 3.

SOS-Based LFC Design for Microgrids

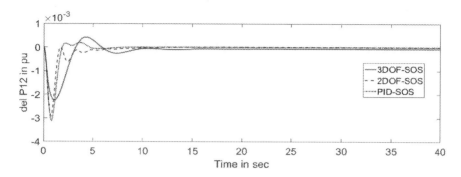

FIGURE 8.14 Deviation of tie line power P_{12}.

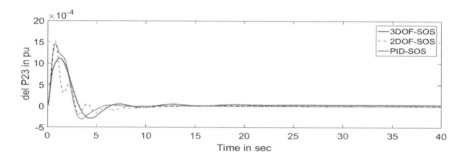

FIGURE 8.15 Deviation of tie line power P_{23}.

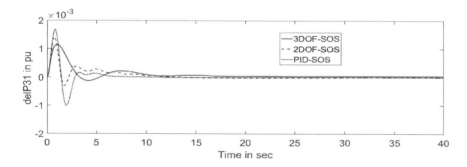

FIGURE 8.16 Deviation of tie line power P_{31}.

Table 8.3 shows the gain parameters of the 3DOF-PID, 2DOF-PID and PID controller optimized by the SOS algorithm. The 3DOF-PID controller has three extra tuning control parameters to enhance the dynamic stability of the system, whereas the others have two controllers. Table 8.4 gives the tabulated data of PID controller parameters optimized by the SOS, TLBO and DEPSO algorithms. A bar diagram

TABLE 8.2
Performance Values of Different Controllers Optimized by SOS Algorithm

Performance Characteristics	Controller	Δf_1 in hz	Δf_2 in hz	Δf_3 in hz	Δf_4 in hz	Δf_5 in hz	Δf_6 in hz
Settling time (T_s) in s	3DOF-PID	4.95	5.20	6.31	7.12	8.14	9.70
	2DOF-PID	5.32	6.07	7.78	7.94	10.28	10.02
	PID	5.92	7.21	8.03	8.57	11.67	12.49
Undershoot (U_{sh}) in pu	3DOF-PID	−0.1982	−0.0424	−0.0445	−2.6098	−0.1854	−0.0692
	2DOF-PID	−0.3634	−0.2899	−0.2515	−5.7181	−1.3212	−1.0542
	PID	−0.4115	−0.3609	−0.3712	−7.1458	−1.5278	−1.6228
Overshoot (O_{sh}) in pu	3DOF-PID	0.0290	0.0166	0.0142	0.2464	0.4500	0.4114
	2DOF-PID	0.1857	0.0740	0.0423	1.6444	3.2698	2.5776
	PID	0.2029	0.0811	0.0721	1.9020	4.5523	3.1612

TABLE 8.3
Gain Parameters of Different Controllers Optimized by SOS Algorithm

Gain parameters	3DOF-PID controller	2DOF-PID controller	PID controller
K_1	0.0100	0.3260	0.7924
K_2	2.0000	2.0000	2.0000
K_3	0.0100	1.2173	0.6479
b_1	0.3315	2.0000	–
c_1	1.8953	1.2492	–
Gf_1	0.0100	–	–
N_1	145.3172	227.5443	–
K_4	2.0000	0.0100	2.0000
K_5	1.6690	2.0000	1.1671
K_6	1.8404	0.9427	1.2854
b_2	0.2721	0.5173	–
c_2	0.0100	2.0000	–
Gf_2	0.1001	–	–
N_2	100.0000	300.0000	–
K_7	2.0000	0.0172	0.4397
K_8	2.0000	1.7371	1.5822
K_9	0.0100	0.9704	1.0167
b_3	0.1026	2.0000	–

Gain parameters	3DOF-PID controller	2DOF-PID controller	PID controller
c_3	0100	0.0100	–
Gf_3	1.1065	–	–
N_3	100.0000	118.4932	–
K_{10}	0100	0.4217	0.1000
K_{11}	1651	1.7643	2.0000
K_{12}	0.4287	1.8982	0.1000
b_4	1.7726	0.4975	–
c_4	0.0100	1.2651	–
Gf_4	1.1079	–	–
N_4	199.4876	218.8336	–

TABLE 8.4
Gain Parameters of PID Controller Tuned by Different Algorithms

Gain Parameters	SOS-PID Controller	TLBO-PID Controller	DEPSO-PID Controller
K_{p1}	1.2086	1.8480	1.2336
K_{i1}	0.0100	0.0447	0.0100
K_{d1}	0.6184	1.4159	1.4782
K_{p2}	0.2616	0.4268	1.5953
K_{i2}	1.7796	0.7509	1.1392
K_{d2}	0.6048	0.7686	1.2427
K_{p3}	0.0100	1.0974	0.4492
K_{i3}	0.0672	1.8517	0.4666
K_{d3}	0.2782	1.5531	0.9432
K_{p4}	0.1269	0.9598	0.7392
K_{i4}	0.4434	0.6480	1.1719
K_{d4}	0.0967	1.0620	1.2311

presentation of the controller's dynamic behaviors is shown in Figures 8.17–8.19. Also, bar diagram presentations of the SOS, TLBO and DEPSO algorithm-based PID controller are shown from Figures 8.20 to 8.21.

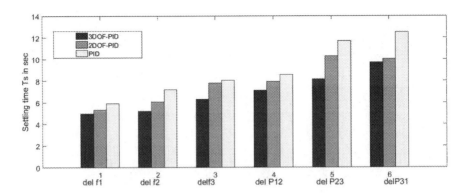

FIGURE 8.17 Comparison of settling time (T_s) for different controllers.

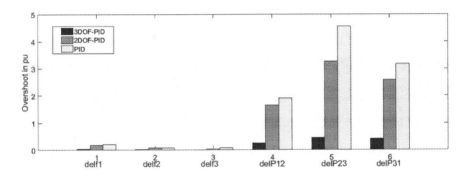

FIGURE 8.18 Comparison of overshoot (O_{sh}) for different controllers.

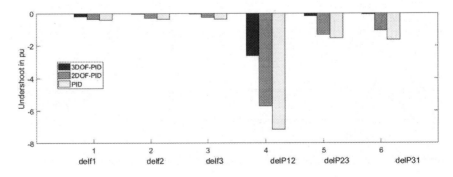

FIGURE 8.19 Comparison of undershoot (U_{sh}) for different controllers.

SOS-Based LFC Design for Microgrids

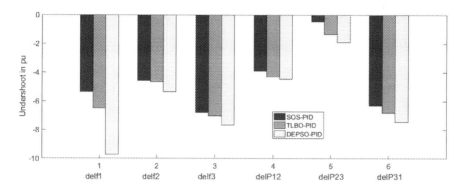

FIGURE 8.20 Comparison of undershoot (U_{sh}) optimized by different algorithms.

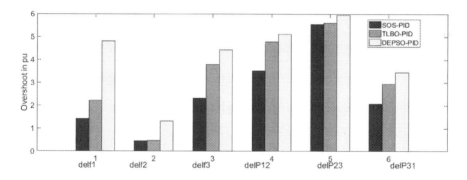

FIGURE 8.21 Comparison of overshoot (O_{sh}) optimized by different algorithms.

8.4 CONCLUSION

In this analysis, load frequency control is accomplished by a hybrid power system comprising conventional thermal, hydro plant and energy resources of microgrid. Analysis of the performances of 3DOF-PID, 2DOF-PID and PID controllers has been given. The influential metaheuristic symbiotic organism search (SOS) technique is used for tuning the gain parameters of 3DOF-PID, 2DOF-PID and PID controllers. The SOS-optimized 3DOF-PID controller delivers greater performance than the others in terms of small overshoots, undershoots and less settling time. The SOS technique performance is superior as compared to TLBO and hybrid DEPSO. The future scope of this work is analysis of the proposed LFC system for deregulated power systems.

8.5 APPENDIX

Subscript refers to area (i) = 1, 2 and 3;
Frequency f = 60 Hz
Turbine time constant of thermal unit (T_{ti}) = 0.3 s;
Reheated turbine time constant and gain are (T_{ri}) and (K_{ri}) = 10 s and 0.5
Gain (K_{pi}) = 120 Hz/pu MW and time constant (T_{pi}) of generator = 20 s
Power interchange among areas (T_{ij} MW/rad) = 0.086 pu
Inertia constant (H_i) = 5 s
Speed regulation parameter (R_i) = 2.4pu Hz/MW
Bias coefficient (B_i) = 0.425 pu MW/Hz
Wind turbine generation gain (K_{WTG}) = 1
Gain of aqua electrolyzer (K_{AE}) = 0.002
Gain of fuel cell (K_{FC}) = 0.01
Diesel engine generator gain (K_{DEG}) = 0.0003
Battery energy storage system gain (K_{BESS}) = −0.0003
Time constant of wind turbine (T_{WTG}) = 1.5 sec
Time constant of AE (T_{AE}) = 0.5 sec
Time constant of FC (T_{FC}) = 4 sec
Time constant of DEG (T_{DEG}) = 2 sec
Time constant of BESS (K_{BESS}) = 0.1 sec
Hydro turbine start time (T_w) = 1 s
Gain of AC/DC link (K_{hvdc}) = 1
Time constant of AC/DC link $\left(T_{hvdc}\right)$ = 0.2 sec
Gains of electric governor K_p, K_i, K_d = 1, 5 and 4.

REFERENCES

1. Concordia, C., Kirchmayer, L. K.: 'Tie-line power and frequency: part II', *AISE Trans.*, III-A, April 1954, 73, pp. 133–146.
2. Wang, Y., Zhou, R., Wen, C.: 'New robust adaptive load-frequency control with system parametric uncertainties', *IEE Proc., Gener. Transm. Distrib.*, 1994, 141, (3), pp. 184–190.
3. Elgerd, O.I., Fosha, C.E.: 'Optimum megawatt–frequency control of multiarea electric energy systems', *IEEE Trans. Power Appar. Syst.*, 1970, PAS-89, (4), pp. 556–563.
4. Debbarma, S., Saikia, L.C., Sinha, N.: 'Solution to automatic generation control problem using firefly algorithm optimized IIDμ controller', *ISA Trans.*, 2014, 53, pp. 358–366.
5. Rahman, A., Saikia, L.C., Sinha, N.: 'AGC of an unequal four-area thermal system using biogeography based optimised 3DOF-PID controller', *IET Renew. Power Gener*, 2016, 10.
6. Kothari, M.L., Nanda, J.: 'Application of optimal control strategy to automatic generation control of a hydrothermal system', *IEE Proc.*, 1988, 135, (4), pp. 268–274.
7. Mohanty, B., Panda, S., Hota, P. K.: 'Controller parameters tuning of differential evolution algorithm and its application to load frequency control of multi source power system', *Electr. Power Energy Syst.*, 2014, 54, pp. 77–85.
8. Tan, W., Chang, S., Zhou, R.: 'Load frequency control of power system with non linearities', *IET, Gen. Trans. Dist.*, October 2017, 11, (17), pp. 4307–4313.
9. Saikia, L.C., Nanda, J., Mishra, S.: 'Performance comparison of several classical controllers in AGC for multi-area interconnected thermal system', *Int. J. Electr. Power Energy Syst.*, 2011, 33, pp. 394–401.

10. Debbarma, S., Saikia, L.C., Sinha, N.: 'Robust two-degree-of-freedom controller for automatic generation control of multi-area system', *Int. J. Elect. Power Energy Syst.*, December 2014, 63, pp. 878–886.
11. Sahu, R.K., Panda, S., Rout, U.K.: 'DE optimized parallel 2-DOF PID controller for load frequency control of power system with governor dead-band nonlinearity', *Int. J. Electr. Power Energy Syst.*, 2013, 49, pp. 19–33.
12. Rahman, A., Saikia, L.C., Sinha, N.: 'AGC of dish-Stirling solar thermal integrated thermal system with biogeography based optimized three degree of freedom PID controller', *IET Renew. Power Gener.*, 2016, 10.
13. Hussain, I., Ranjan, S., Das, D.C., et al.: 'Performance analysis of flower pollination algorithm optimized PID controller for wind-PV-SMES-BESS diesel autonomous hybrid power system', *Renew. Energy Res.*, 2017, 7, (2), pp. 643–651.
14. Pandey, S.K., Mohanty, S.R., Kishor, N., et al.: 'Frequency regulation in hybrid power systems using particle swarm optimization and linear matrix inequalities based robust controller design', *Electr. Power Energy Syst.*, 2014, 63, pp. 887–900.
15. Saha, D., Saikia, L.C.: 'Performance of FACTS and energy storage devices in a multi area wind-hydro-thermal system employed with SFS optimized I-PDF controller', *Renew. Sustain. Energy*, 2017, 9, p. 024103.
16. Raju, M., Saikia, L.C., Sinha, N.: 'Load frequency control of multi-area hybrid power system using symbiotic organisms search optimized two degree of freedom controller', *Renew. Energy Res.*, 2017, 7, (4), pp. 1663–1674.
17. Mirazimi, J., Fathi, M.: 'Analysis of hybrid wind/fuel cell/battery/diesel energy system under Alaska condition', 8th Electrical Electronics Computer Telecommunications Information Technology Association (ECTI), Thailand, 2011, pp. 917–92.
18. Howlader, A.M., Izumi, Y., Uehara, A., Urasaki, N., Senjyu, T., Yona, A., Saber, A.Y.: 'A minimal order observer based frequency control strategy for an integrated windbatterydiesel power system', *Energy*, October 2012, 46, (1), pp. 168–178.
19. Vidyanandan, K.V., Senroy, N.: 'Frequency regulation in a wind–diesel powered microgrid using flywheels and Fuel Cell', *IET Gener. Trans. Distr.*, 2016, 10, (3), pp. 780–788.
20. Serban, I., Marinescu, C.: 'Battery energy storage system for frequency support in microgrids and with enhanced control features for uninterruptible supply of local loads', *Electr. Power Energy Syst.*, January 2015, 54, pp. 432–441.
21. Alexandre Oudalov, E., Chartouni, D., Ohler, C.: 'Optimizing a battery energy storage system for primary frequency control', *IEEE Trans. Power Syst.*, 2007, 22.
22. Nanda, J., Mangla, A., Suri, S.: 'Some new findings on automatic generation control of an interconnected hydrothermal system with conventional controllers', *IEEE Trans. Energy Convers.*, 2006, 21, (1), pp. 187–194.
23. Khuntia, S.R., Panda, S.: 'Simulation study for automatic generation control of a multi-area power system by ANFIS approach', *Appl. Soft Comput.*, 2012, 12, pp. 333–341.
24. Guha, D., Roy, P.K., Banerjee, S.: 'Load frequency control of interconnected power system using grey wolf optimization', *Swarm Evol. Comput.*, 2016, 27, pp. 97–115.
25. Ali, E.S., Abd-Elazim, S.M.: 'BFOA based design of PID controller for two areas load frequency control with nonlinearities', *Int. J. Electr. Power Energy Syst.*, 2013, 51, pp. 224–231.
26. Sahu, Binod Kumar, et al.: 'Teaching–learning based optimization algorithm based fuzzy-PID controller for automatic generation control of multi-area power system', *Appl. Soft Comput.*, 2015, 27, pp. 240–249.
27. Pahadasingh, S., Jena, C., Panigrahi, C.K., 'Fuzzy PID AGC of multi area power system optimized by hybrid DEPSO algorithm with FACTS', *Inter. J. Recent Technol. Eng.*, 2019, 8, pp. 9563–9569.
28. Nayak, R.J., Sahu, K.B.: 'Load frequency control of hydro thermal system using fuzzy PID controller optimized by hybrid DECPSO algorithm', *Inter. J. Pure Appl. Math.*, August 2017, 114, pp. 147–155.

29. Patel, N.C., Debnath, M.K., Bagarty, D.P., Das, P.: 'GWO tuned multi degree of freedom PID controller for LFC', *Inte. J. Eng. Technol.*, 2018, 7, pp. 548–552.
30. Debbarma, S., Saikia, L.C., Sinha, N.: 'Automatic generation control using two degree of freedom fractional order PID controller', *Int. J. Electr. Power Energy Syst.*, 2014, 58, pp. 120–129.
31. Raju, M., Saikia, L.C., Sinha, N.: 'Load frequency control of multi-area hybrid power system using symbiotic organisms search optimized two degree of freedom controller', *Renew. Energy Res.*, 2017, 7, (4), pp. 1663–1674.
32. Singh, J., Chatterjee, K., Vishwakarma, C.B.: 'Two degree of freedom internal model control-PID design for LFC of power systems via logarithmic approximations', *ISA Trans.*, 2018, 72, pp. 185–196.
33. Dash, P., Saikia, L.C., Sinha, N.: 'Comparison of performances of several FACTS devices using cuckoo search algorithm optimized 2DOF controllers in multi-area AGC', *Electr. Power Energy Syst.*, 2015, 1, (65), pp. 316–324.
34. Debnath, M.K., Satapathy, P., Mallick, R.K.: '3DOF-PID controller based AGC using TLBO algorithm', *Inter. J. Pure Appl. Math.*, 2017, 114, pp. 39–49.
35. Rahman, A., Saikia, L.C., Sinha, N.: 'Load frequency control of a hydro-thermal system under deregulated environment using biogeography-based optimized three-degree-of-freedom integral–derivative controller', *IET Gener. Trans. Distr.*, 2015, 9, (15), pp. 2284–2293.
36. Ibraheem, N., Bhatti, T.S.: 'AGC of two area power system interconnected by AC/DC links with diverse sources in each area', *Electr. Power Energy Syst.*, 2014, 55, pp. 297–304.
37. Barisal, A.K.: 'Comparative performance analysis of teaching learning based optimization for automatic load frequency control of multi-source power systems', *Electr. Power Energy Syst.*, 2015, 66, pp. 67–77.
38. Arya, Y., Kumar, N., Ibraheem: 'AGC of a two-area multi-source power system interconnected via AC/DC parallel links under restructured power environment', *Optim. Control Appl. Methods.*, 2016, 37, pp. 590–607.
39. Arya, Y., Kumar, N.: 'AGC of a multi-area multi-source hydrothermal power system interconnected via AC/DC parallel links under deregulated environment', *Electr. Power Energy Syst.*, 2016, 75, pp. 127–138.
40. Rao, R.V., Savsani, V.J., Vakharia, D.P.: 'Teaching–learning-based optimization: An optimization method for continuous non-linear large scale problems', *Inf. Sci.*, 2012, 183, (1), pp. 1–15.
41. Guha, D., Roy, P.K., Banerjee, S.: 'Maiden application of SSA optimized CC-TID controller for LFC of power systems', *IET Gener. Trans. Distr.*, 2019, 13, pp. 1110–1120.
42. Raju, M., Saikia, L.C., Sinha, N.: 'Load frequency control of a multi-area system incorporating distributed generation resources, gate controlled series capacitor along with high-voltage direct current link using hybrid ALO-pattern search optimised fractional order controller', *IET Renew. Power Gener.*, 2019, 13, (2), pp. 330–341.
43. Cheng, M., Prayogo, D.: 'Symbiotic organisms search: A new metaheuristic optimization algorithm', *Inter. J. Comp. Struc.*, 2014, 139, pp. 98–112.
44. Guha, D., Roy, P.K., Banerjee, S.: 'Symbiotic organism search based load frequency control with TCSC', Proc. Fourth Int. Conf. Recent Advances in Information Technology, IIT (ISM) Dhanbad, 15–17 March 2018, pp. 1–6.
45. Kundur, P.: *Power system stability and control* (McGraw Hill, New York, 1994, 2nd edn.).

9 Performance Improvement of a UPQC Integrated with a Microgrid System Using Modified SRF Technique

Sarita Samal, Prasanta Kumar Barik, and Prakash Kumar Hota

CONTENTS

9.1 Introduction	201
9.2 Proposed System	203
9.2.1 Modeling of Solar PV	203
9.2.2 Fuel Cell Modeling	205
9.3 UPQC Detail Model	207
9.3.1 Control Scheme of UPQC	209
9.3.1.1 SRF Control Scheme	209
9.3.1.2 MSRF Control Scheme	210
9.4 Results and Discussion	210
9.4.1 Scenario-1 Analysis	211
9.4.2 Scenario-2 Analysis	212
9.5 Conclusion	212
References	215

9.1 INTRODUCTION

Microgrid (MG) may be represented as a small-scale power system that contains loads, energy sources, energy storage units and control and protection systems [1]. Using an MG is attractive, as it improves the system quality, decreases carbon emissions and reduces losses in transmission and distribution systems [2]. MGs can be connected to the main grid or operate autonomously. When an MG is connected to utility grid, the control systems required to maintain the active and reactive power output from the energy sources connected to the MG are simple. However, under autonomous operation, the MG is disconnected from the utility grid and operates in islanded mode [3, 4]. Usually, a stand-alone MG system is used to supply power to isolated areas or places interconnected to a weak grid. The application of an MG reduces the probability of energy supply scarcity. The proposed MG consists of renewable energy source-based power sources, that is, solar PV and a fuel cell-based MG. Solar PV and wind energy are complementary in nature, and both depends upon climatic conditions; hence, to get an uninterrupted power supply at any time and maintain the continuity of the load current, one of the most-developed energy sources, like a fuel cell, is combined with these RESs [5]. However, the electric power system is mostly affected by nonlinear loads; arc furnaces, switch mode power supply, power electronic converters and household electronic equipment play a key role in polluting supply voltages and currents. The increase of power electronic-based equipment in household appliances and industries is the main cause of pollution of power systems [6]. Consequently, power quality (PQ) improvement is a major issue today. Research in the area of power electronics makes sure that unified power quality conditioner (UPQC) plays a vital role for achieving superior PQ levels. In the present scenario, the series active power filter (APF) and shunt APF alone do not meet the requirements for compensating PQ distortions. However, the UPQC is a device which consists of these two APFs integrated with the DC-link capacitor. The series APF is integrated though a series transformer and the shunt APF (SAPF) through an interfacing inductor. The series inverter acts as a voltage source, whereas the shunt acts as a current source. Simultaneous compensation of voltage and current-related PQ distortions using UPQC is achieved by proper control of series APF and SAPF. The SAPF is employed to provide compensating currents to the PCC for generation/absorption of reactive power and harmonics suppression. Moreover, the operation of the SAPF depends upon three main parts which are important in its design; these consist of the control method used for generation of the reference current, the technique used for switching pulse generation for the inverter and the controller used for DC link capacitor voltage regulation. Different control strategies explained in literature are as follows. The typical use of SAPFs for current harmonic compensation in domestic, commercial and industrial applications was explained in Hosseinpour et al. [7]. An experimental study and simulation design of an SAPF for harmonics and reactive power compensation were explained by Montero et al. [8]. The power balance theory for active and reactive power compensation was developed by Jain et al. [9]. The instantaneous reactive power techniques of three-phase SAPF for compensation of source current harmonics were explained by Singh et al. [10]. Sag is the most significant PQ problem faced by many industrial consumers. The control for such a case can be analyzed by protecting sensitive loads in order to preserve a load voltage without sudden phase shifts [11]. Different control strategies for series APF were analyzed by Ferdi et al. [12] with a focus on the reimbursement of voltage sags with phase jump. Different control techniques to reimburse voltage sags with phase jump were also examined and compared by Jowder et al. [13]. To ensure stable operation and improve the system performance of distributed generation in islanded mode, a comparative study of two different control techniques used in UPQC-like reference current generation, that is, the synchronous reference frame (SRF) and modified synchronous reference frame (MSRF) methods in conjunction with a pulse width modulation-based hysteresis band controller is proposed in this chapter using MATLAB simulation software. The PQ issues like voltage sag compensation, current and voltage harmonics are analyzed both in linear and nonlinear loads.

9.2 PROPOSED SYSTEM

The projected MG system (comprising solar and fuel cell-based energy sources) is shown in Figure 9.1, where the MG system generates DC power to the DC bus, and by using a power inverter, this DC power is converted to AC. The AC bus delivers the power to the load, which may be a linear or nonlinear. The UPQC is located between the MG and nonlinear load, which manages the power quality of the system by using different control techniques.

9.2.1 MODELING OF SOLAR PV

A single diode model-based PV cell is used for the design of the MG. Figure 9.2 represents the single-diode equivalent model of the solar PV system. The basic equation for the design of the PV system is given in the following [14, 15].

$$I = I_s - I_d - I_{sh} \tag{9.1}$$

$$I_S = \left[I_{SC} + K_i\left(T_k T\right)\right] \times \frac{G}{1000} \tag{9.2}$$

$$I_{RS} \frac{I_{sc}}{\left[\exp\left(q \times V_{OC} / N_S \times k \times A \times T\right) - 1\right]} \tag{9.3}$$

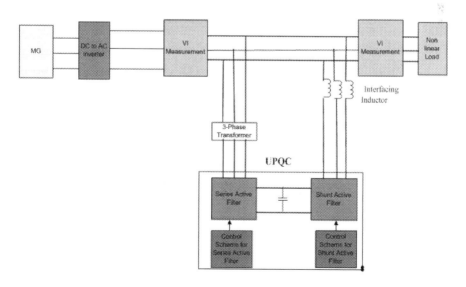

FIGURE 9.1 Proposed MG and UPQC.

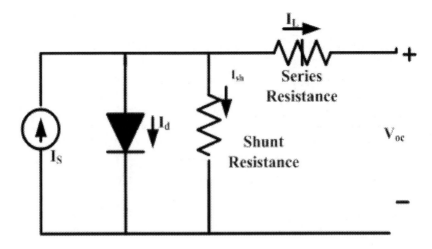

FIGURE 9.2 SPV model.

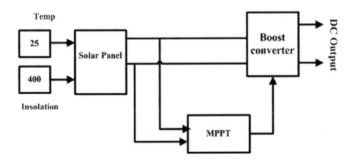

FIGURE 9.3 SPV with boost converter and MPPT.

$$I_o = I_{RS} \left[\frac{T}{T_r}\right]^3 \exp\left[\frac{q \times E_{go}}{Ak}\left\{\frac{1}{T_r} - \frac{1}{T}\right\}\right] \quad (9.4)$$

$$I_{PV} = N_P \times I_{Ph} - N_P \times I_O \left[\exp\left\{\frac{q \times V_{PV} + I_{PV} \times R_{se}}{N_s \times AkT}\right\} - 1\right] \quad (9.5)$$

where R_{se} = resistance in series (Ω), I_0 = reverse saturation current of diode (A), V_{oc} = open circuit voltage (V), R_{sh} = shunt resistance (Ω), I_{sc} = short circuit current (A), I_{PV} = diode photo current (A), V_{PV} = diode voltage (V), k = Boltzmann constant, N_P = parallel connected cells, T = temperature of the p-n junction (Kelvin), N_s = cells in series, A = diode ideality factor, q = electron charge.

Based on Equations (9.1–9.5), a single-diode model solar PV cell is developed and implemented in MATLAB simulations [16–20]. Figure 9.3 shows the MATLAB simulation of PV with an MPPT and boost converter. Figure 9.4 shows its corresponding

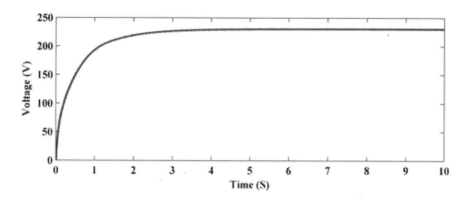

FIGURE 9.4 Voltage output.

TABLE 9.1
Different Parameters of Solar PV

Different Parameters	Ratings
N_p	72
N_s	01
I_{sc}	10.2 A
V_{oc}	90.5 V
V_{mp}	81.5 V
I_{mp}	8.6 A
V_0	230 V

output voltage, where the required voltage of 230 V is achieved. The parameters required for design of the solar PV system are given in Table 9.1.

9.2.2 Fuel Cell Modeling

The proton exchange membrane (PEM) FC is considered another energy source of the MG. The FC consists of two electrodes, a positive cathode and negative anode, and an electrolyte. The pressurized hydrogen gas enters as the anode of the FC and oxygen enters the cathode [21, 22]. A basic PEMFC diagram is shown in Figure 9.5, and its chemical reactions are presented in Equations (9.6–9.8) [23].

The reaction at the anode is given as:

$$(2H^+ + 2e^-) \rightarrow H_2 \tag{9.6}$$

The reaction at the cathode is given as:

$$(2H^+ + 2e^- + 1/2O_2) \rightarrow H_2O \tag{9.7}$$

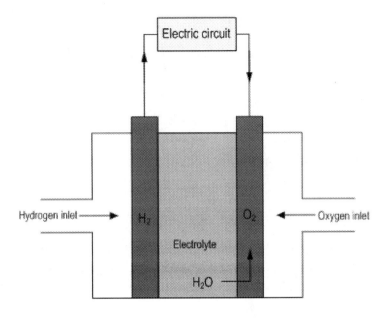

FIGURE 9.5 FC model.

The net reaction is shown as:

$$H_2 + 1/2O_2 \rightarrow H_2O \tag{9.8}$$

$$E_{Thermo} = 1.229 - 0.00085 \times (T - 298.15) + 4.31 \times 10^{-5} \times T \\ \times \left[\ln(P_{H2}) + \frac{1}{2}\ln(P_{O2}) \right] \tag{9.9}$$

$$V_{act} = -\left[\xi 1 + \xi 2 \times T + \xi 3 \times T \times \ln(CO_2) \right] \tag{9.10}$$

$$V_{Ohmic} = i_{FC}(R_M + R_C) \tag{9.11}$$

$$V_{FC} = E_N - V_{act} - V_{Ohmic} - V_{con} \tag{9.12}$$

$$V_s = k \times V_{FC} \tag{9.13}$$

$$V_{con} = \ln\left(1 - \frac{J}{J_{max}}\right) \times (-B) \tag{9.14}$$

The thermodynamically predicted voltage (E_{thermo}) of FC is shown in Equation (9.10) [24]. Equations (9.9)–(9.14) are required for the modeling of PEMFC in MATLAB [25]. Figures 9.6 and 9.7 represent the detailed model with output voltage. The simulation parameters of FC are presented in Table 9.2.

Performance Improvement Using Modified SRF Technique

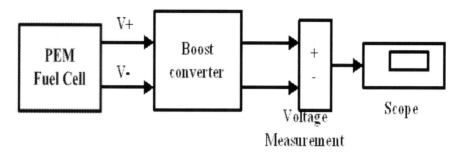

FIGURE 9.6 FC with BC block diagram.

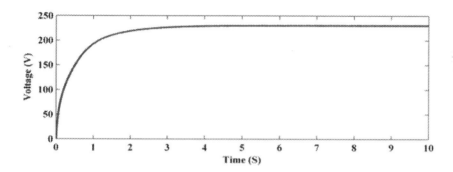

FIGURE 9.7 Voltage output.

TABLE 9.2
FC Parameters

Parameters	Values
O_2	59.3%
V_{fc}	230 V
v_s	1.128 V
H_2	99.56%
k	65
R_t	0.70833 Ω
R_l	5 Ω

9.3 UPQC DETAIL MODEL

This section gives the system configuration and detailed description of the UPQC. The basic structure of the UPQC is shown in Figure 9.8, and it consists of two inverters connected to a common DC-link capacitor. The series inverter is connected

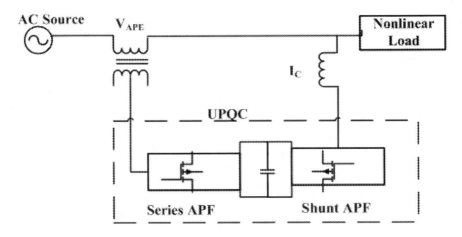

FIGURE 9.8 Basic UPQC model.

through a series transformer, and the shunt inverter is connected in parallel with the point of common coupling. The series inverter acts as a voltage source, whereas the shunt acts as a current source. The main function of the UPQC is to control the power flow and reduce the harmonics distortion both in voltage and current waveforms. The series APF protects the load from utility-side disturbances. In the case of series APF, Park's transformation method is used for generation of the unit vector signal. A PWM generator, generating synchronized switching pulses, is given to the six switches of the series converter. The objectives of the series converter can be achieved by providing appropriate switching pulses.

The shunt active power filter injects the compensating current to the PCC such that the load current becomes harmonics free. With the aid of the voltage source inverter and interfacing inductor, the compensating current is generated, which minimizes the harmonics component in the load current. The maximum *di/dt* that can be accomplished by the inverter is determined by the voltage across the interfacing inductor. This is significant because a higher *di/dt* is essential to reimburse the harmonics; hence, the selection of the interfacing inductor is vital. However, the ability of the compensator depends upon the interfacing inductor, where a higher size can compensate higher-order harmonics. The SAPF generates a compensating current, which is in opposition to the harmonic current generated by the nonlinear load. This compensating current cancels out the current harmonics and makes the load current sinusoidal. So, the SAPF is used to eradicate current harmonics and reimburse reactive power at the source side so as to make the load current harmonics free. Equations (9.15) and (9.16) show the instantaneous current and the source voltage, respectively.

$$I_c(t) = I_L(t) - I_s(t) \tag{9.15}$$

$$V_s(t) = V_m \sin \omega t \tag{9.16}$$

Performance Improvement Using Modified SRF Technique

$$I_s(t) = I_1 \sin(\omega t + \Phi_1) + \sum_{n=2}^{\varepsilon} I_n \sin(n\omega t + \Phi_n) \quad (9.17)$$

$$I_s(t) = I_L(t) - I_c(t) \quad (9.18)$$

The instantaneous value of the source, load and compensation current can be expressed by $I_s(t)$, $I_L(t)$ and $I_c(t)$, where $V_s(t)$ and V_m correspond to the instantaneous value and peak value of the source voltage.

9.3.1 Control Scheme of UPQC

There are many control methods available for SAPF. Here, the following two basic design methods, which are easy to implement, are considered.

- Synchronous reference frame method
- Modified-SRF method

9.3.1.1 SRF Control Scheme

The normal arrangement of an SRF system contains a phase-locked loop (PLL) unit for vector orientation, as shown in Figure 9.9.

The control pattern includes the transfer of source current from *abc* to *d-q*. The three-phase load currents I_{la}, I_{lb}, I_{lc} are converted to I_d–I_q using the transformation technique, as given in Equation (9.19).

$$\begin{bmatrix} i_q \\ i_d \\ i_o \end{bmatrix} = \frac{2}{3} \begin{bmatrix} \cos\theta & \cos(\theta-120) & \cos(\theta+120) \\ \sin\theta & \sin(\theta-120) & \sin(\theta+120) \\ \frac{1}{2} & \frac{1}{2} & \frac{1}{2} \end{bmatrix} \begin{bmatrix} i_{la} \\ i_{lb} \\ i_{lc} \end{bmatrix} \quad (9.19)$$

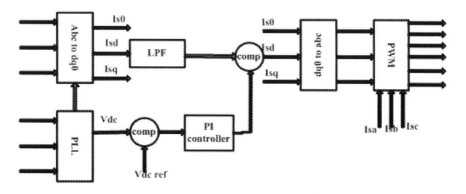

FIGURE 9.9 Block diagram of SRF control scheme.

9.3.1.2 MSRF Control Scheme

Figure 9.10 shows the block diagram of the MSRF scheme, where, instead of PLL, a unit vector generation method is used.

Figure 9.11 shows a block diagram to generate the unit vector by sensing the supply voltage.

$$\cos\theta = \frac{V_\alpha}{\sqrt{(V_{s\alpha}2)+(V_{s\beta}2)}} \quad (9.20)$$

$$\sin\theta = \frac{V_\beta}{\sqrt{(V_{s\alpha}2)+(V_{s\beta}2)}} \quad (9.21)$$

9.4 RESULTS AND DISCUSSION

The proposed model is tested based on following scenarios:

Scenario-1: MG connected to nonlinear load without UPQC and UPQC with SRF technique.
Scenario-2: MG with MSRF-based UPQC.

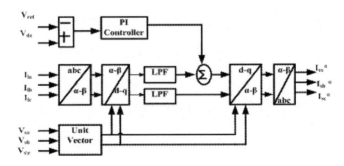

FIGURE 9.10 MSRF scheme.

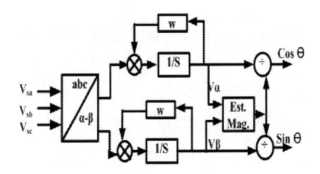

FIGURE 9.11 Unit vector generation model.

9.4.1 Scenario-1 Analysis

In this case, the system performance is analyzed by connecting a non-linear load with the MG system first without UPQC and then with SRF-based UPQC.

The performance of series APF can be evaluated by introducing voltage sag into the system. The profile of load voltage shown in Figure 9.12(a) confirms that voltage sag is introduced from 0.1 to 0.3 s of the load voltage waveform. For the sag condition, the series APF detects the voltage drop and injects the required voltage through the series coupling transformer. It maintains the rated voltage across the load

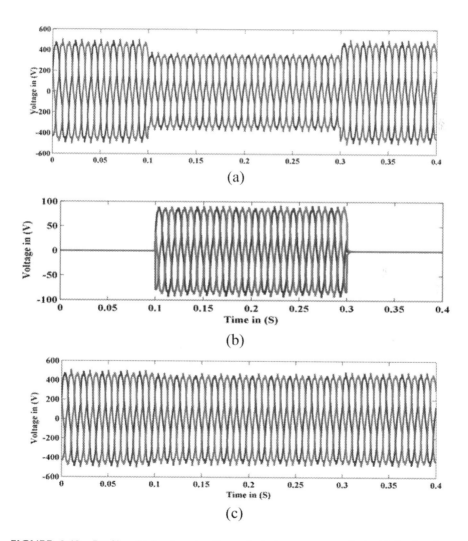

FIGURE 9.12 Profile obtained under Scenario-1 (sag compensation): (a) load voltage before compensation, (b) compensating voltage injected by UPQC and (c) load voltage after compensation.

terminal. In order to compensate the load voltage sag, UPQC (employing an SRF scheme) is turned on, which injects compensating voltage at the PCC, as displayed in Figure 9.12(b); as a result, the load voltage is same as the source voltage. The load voltage after compensation is shown in Figure 9.12(c). In general, the operation of the series part of the UPQC can be described as rapid detection of voltage variations at the source, and it injects the compensation voltage, which maintains the rated voltage across the load terminal.

The shunt voltage source inverter in the UPQC is realized as a shunt APF and is applied to solve the current-related PQ distortion, current harmonic distortion, reactive power demand and so on. In order to investigate the performance of the shunt APF, a rectifier-based nonlinear load is introduced into the system, and the level of harmonics is checked. It is observed from Figure 9.13(a) that the source current waveform has a total harmonic distortion (THD) of 16.60%, per the fast Fourier transform (FFT) analysis of the source current, as shown in Figure 9.13(b). In order to make the source current sinusoidal, the shunt APF of the UPQC with the conventional SRF technique is turned on at $t = 0.1$ s, which injects a compensating current as displayed in Figure 9.13(c). Hence, the THD level comes down to 3.40%, as shown in Figure 9.13(d).

9.4.2 Scenario-2 Analysis

It is observed from Scenario-1 that an MG connected with the SRF method mitigates the harmonics only by 3.40%. Hence, the PQ analysis of the MG system by utilizing the MSRF scheme is studied in this case. As the main objective of the research is to compensate sag, decrease harmonics and maintain the DC link capacitor voltage, the investigation is focused on series APF and shunt APF of the UPQC.

The voltage injected by the proposed controller, shown in Figure 9.14(a), reduces the sag from 0.1 to 0.3 s and makes the load voltage equal to the source voltage. The load voltage after compensation is presented in Figure 9.14(b). The profile of the compensating current generated by the MSRF-based shunt APF and source current after compensation are shown in Figures 9.14(c) and 9.14(d), respectively. It may be seen from Figure 9.14(d) that the source current is sinusoidal in nature due to the injection of compensating current by the UPQC. FFT analysis of the source current (after compensation) is presented in Figure 9.14(e) and ensures further reduction of THD content to 2.54% in comparison to Scenario-1. The simulation is carried out for a non-linear load with these controllers and makes the source current almost sinusoidal after compensation. FFT analysis of the proposed method confirms that the THD of the source current is in compliance with IEEE-519 harmonic standards.

9.5 CONCLUSION

The investigation shows that that MSRF technique of UPQC makes it possible to lighten the PQ of a MG system connected with a nonlinear load. Moreover, the proposed system also enhances the performance of the UPQC by attaining a lower THD value of the source current and voltage sag compensation. From the numerical comparison of results, it is noticeable that the MSRF-based UPQC can reduce the source

Performance Improvement Using Modified SRF Technique 213

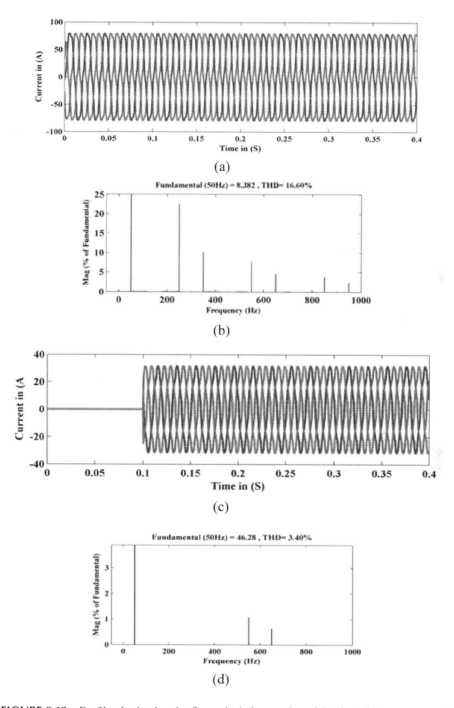

FIGURE 9.13 Profile obtained under Scenario-1 (harmonics mitigation): (a) source current before compensation, (b) harmonics content before compensation, (c) compensating current injected by UPQC and (d) harmonics content after compensation.

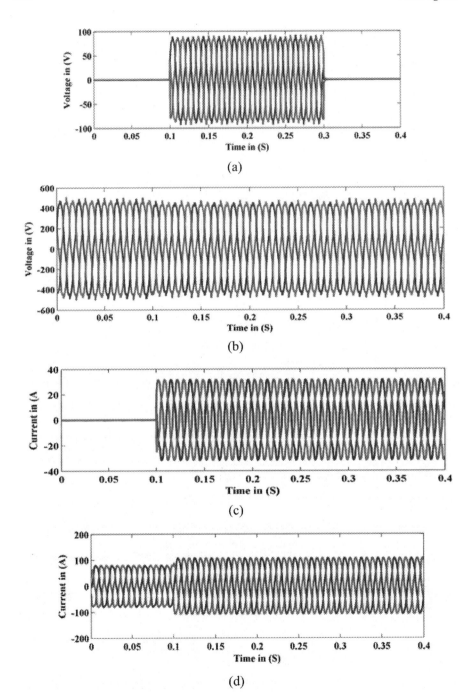

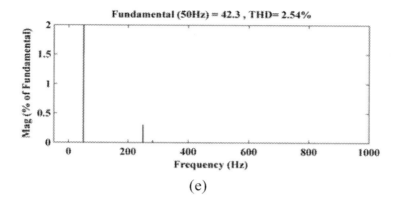

(e)

FIGURE 9.14 Profile obtained under Scenario-2: (a) compensating voltage injected by series APF of UPQC, (b) load voltage after compensation, (c) compensating current injected by shunt APF of UPQC, (d) source current after compensation and (e) harmonics content after compensation.

current harmonics about 2.54%. Investigations may be made into using a hybrid filter and designing its control loop in order to improve the power quality of the system in various aspects.

REFERENCES

1. Badoni, M.; Singh, B.; Singh, A.: Implementation of echo-state network-based control for power quality improvement, *IEEE Transactions on Industrial Electronics*, 2017, Vol.64(7), pp. 5576–5584.
2. Rabiee, A.; Sadeghi, M.; Aghaeic, J.; Heidari, A.: Optimal operation of microgrids through simultaneous scheduling of electrical vehicles and responsive loads considering wind and PV units uncertainties, *Renewable and Sustainable Energy Reviews*, 2016, Vol.57, pp. 721–739.
3. Mahmoud, M.S.; Rahman, M.S.U.; Fouad, M.S.: Review of microgrid architectures—a system of systems perspective, *IET Renewable Power Generation*, 2015, Vol.9(8), pp. 1064–1078.
4. Samal, S.; Hota, P.K.: Design and analysis of solar PV-fuel cell and wind energy based microgrid system for power quality improvement, *Cogent Engineering*, 2017, Vol.4(1), pp. 1402453.
5. Suresh, M.; Patnaik, S.S.; Suresh, Y.; Panda, A.K.: Comparison of two compensation control strategies for shunt active power filter in three-phase four-wire system, Innovative Smart Grid Technologies (ISGT), IEEE PES, 2011, pp. 1–6.
6. Tang, Y.; Loh, P.C.; Wang, P.; Choo, F.H.; Gao, F.; Blaabjerg, F.: Generalized design of high performance shunt active power filter with output LCL filter, *IEEE Transactions on Industrial Electronics*, 2012, Vol.59(3), pp. 1443–1452.

7. Hosseinpour, M.; Yazdian, A.; Mohamadian, M; Kazempour, J.: Design and simulation of UPQC to improve power quality and transfer wind energy to grid, *Journal of Applied Sciences*, 2008, Vol.8(21), pp. 3770–3782.
8. Montero, M.I.M.; Cadaval, E.R.; Gonzalez, F.B.: Comparison of control strategies for shunt active power filters in three-phase four-wire systems, *IEEE Transactions on Power Electronics*, 2007, Vol.22(1), pp. 229–236.
9. Samal, S.; Hota, P.K.; Barik, P.K.: Performance improvement of a distributed generation system using unified power quality conditioner. *Technology and Economics of Smart Grids and Sustainable Energy*, 2020, Vol.5(1), pp. 1–16.
10. Singh, B.N.; Singh, B.; Chandra, A.; Al-Haddad, K.: Design and digital implementation of active filter with power balance theory, *IEE Proceedings-Electrical Power Application*, 2005, Vol.152(5), pp. 1149–1160.
11. Dixon, J.W.; Venegas, G.; Moran, L.A.: A series active power filter based on a sinusoidal current-controlled voltage-source inverter, *IEEE Transactions on Industrial Electronics*, 1997, Vol.44(5), pp. 612–620.
12. Ferdi, B.; Benachaiba, C.; Dib, S.; Dehini, R.: Adaptive PI control of dynamic voltage restorer using fuzzy logic, *Journal of Electrical Engineering: Theory & Application*, 2010, Vol.1(3), pp. 165–175.
13. Jowder, F.A.L.: Design and analysis of dynamic voltage restorer for deep voltage sag and harmonic compensation, *IET Generation, Transmission & Distribution*, 2009, Vol.3(6), pp. 547–560.
14. Akagi, H.; Kanazawa, Y.; Nabae, A.: Instantaneous reactive power compensators comprising switching devices without energy storage components, *IEEE Transactions on Industrial Electronics Application*, 1984, Vol.20(3), pp. 625–630.
15. Barik, P.K.; Shankar, G.; Sahoo, P.K.: Power quality assessment of microgrid using fuzzy controller aided modified SRF based designed SAPF, *International Transactions on Electrical Energy Systems*, 2019, p. e12289. https://doi.org/10.1002/2050-7038.12289.
16. Hatziargyriou, C.S.; Liang, T.: *The microgrids concept, microgrid: Architectures and control*, John Wiley-IEEE Press, Chichester, 2014, pp. 1–24.
17. Olivares, D.E., et al.: Trends in microgrid control, *IEEE Transactions on Smart Grid*, 2014, Vol.5(4), pp. 1905–1919.
18. Altas, I.H.; Sharaf, A.M.: A photovoltaic array simulation model for MATLAB-Simulink GUI environment, Clean Electrical Power, ICCEP'07, pp. 341–345.
19. Femia, N.; Petrone, G.; Spagnuolo, G.; Vitelli, M.: Optimization of perturb and observe maximum power point tracking method, *IEEE Transactions on Power Electronic*, 2007, Vol.20(4), pp. 963–973.
20. Samal, S.; Hota, P.K.; Barik, P.K.: Harmonics mitigation by using shunt active power filter under different load condition, 2016 International Conference on Signal Processing, Communication, Power and Embedded System (SCOPES), October 2016, pp. 94–98.
21. Kalirasu, A.; Dash, S.S.: Simulation of closed loop controlled boost converter for solar installation, *Serbian Journal of Electrical Engineering*, 2010, Vol.7(1), pp. 121–130.
22. Noroozian, R.; Abedi, M.; Gharehpetian, G.B.; Bayat, A.: On-grid and off-grid operation of multi-input single-output DC/DC converter based fuel cell generation system, Electrical Engineering (ICEE), 18th Iranian Conference, 2010, pp. 753–758.
23. Samal, S.; Hota, A.; Hota, P.K.; Barik, P.K.: Harmonics and voltage sag compensation of a solar PV-based distributed generation using MSRF-based UPQC, Innovation in Electrical Power Engineering, Communication, and Computing Technology, Lecture Notes in Electrical Engineering, vol. 630, Springer, 2020.

10 Application of Probabilistic Neural Network and Wavelet Analysis to Classify Power Quality

Pampa Sinha and Chitralekha Jena

CONTENTS

10.1	Introduction	217
10.2	Proposed Method	218
10.3	System Description	220
10.4	Wavelet Transform-Based Feature Extraction	221
	10.4.1 Capacitor Switching	221
	10.4.2 Motor Switching	222
	10.4.3 Fault-Induced Events	223
	10.4.4 Converter Switching	224
	10.4.5 Transformer Switching	225
10.5	Categorization of Transients Using PNN	226
10.6	Comparative Analysis	228
10.7	Conclusions	229
References		229

10.1 INTRODUCTION

One of the approaches used to identify the source of power quality problems in power systems is to use classifiers. Classifiers utilize the characterizing features of various forms of power system signals in order to determine the different types and disturbances. Of the various forms of classifier tools, neural networks have been used for some time. In this chapter, the use of a specific variety of neural network, a probabilistic neural network (PNN), is investigated for the categorization and identification of power quality problems. Transients in power systems are a challenge for the coordination of equipment insulation. Transients occurring due to different

DOI: 10.1201/9781003121626-10

types of switching have typical characterizing frequencies. Wavelet transform can be effectively employed to extract the features contained in these transient signals. In this chapter, a neural network model based on wavelet transform was designed for transient classification due to capacitor, motor, converter, transformer switching or fault. For these five transients, detailed aspects of reactive powers are found, and the proposed model uses reactive power as input to the PNN, which is used to categorize the different transients. In this chapter, an 11 kV distribution system is considered. The transient signals are successfully classified using a neural network.

Various combinations of feature extraction and classification methods have been reported for identification of different problems related to power quality. In [1], wavelet transform was employed as a means for supervising the problems generated due to the dynamic performance of industrial plants. The authors of [2] proposed a wavelet-based NN model for identifying power quality problems, which was executed and tested with a number of transient events. For developing a classifier, discrete wavelet transform (DWT) is used with the PNN model. In [3], the authors proposed a method to distinguish transients arising. A DWT of the nodal voltage signal was applied to find typical properties of the voltage waveform. The authors of [4] presented a new approach for meticulous differentiation between magnetizing inrush current and internal fault in a power transformer by considering wavelet transforms with NNs. The authors of [5] employed sparse representation classification (SRC) in addition to a feature extraction procedure using dual-tree complex wavelet transform (DTCWT). The authors of [6] proposed a DTCWT for identification of power system transients. An S-transform-based competitive neural network (CNN) classifier was described in [7] for recognition of inrush current.

The authors of [8] reviewed the literature for power quality on existing applications of advanced artificial intelligence approaches. It includes an extended accumulation of literature considering fuzzy logic applications and neural networks as well as genetic algorithms in power quality. A PNN is utilized for classifying transients related to power systems. In experiments for power transients, it has been shown that the classification system achieves a 99% classification rate and is robust in noisy environments. A method of utilizing PNN in the dynamic security assessment of power systems was proposed in [9].

The authors of this chapter presented different transients like motor or capacitor switching, various fault-induced transients and transformer and converter inrush transients utilizing PNN approaches. To find the exceptional characteristic of each transient, first, the DWT of the signal is utilized for the calculation of the reactive power at levels 1 and 3. As the ANN has the possibility to simulate the learning process as well as working as a classifier, a PNN is used for classifying the stated disturbances in transients, and it is verified that this single-point approach is very efficient for classification of transient signals.

10.2 PROPOSED METHOD

A novel method is proposed in this chapter, which is based on PNN and DWT for categorization of motor, converter, capacitor or transformer switching as well as fault-induced events. A flowchart is shown in Figure 10.1 to describe the proposed method

Application of PNN and Wavelet Analysis to Classify Power Quality

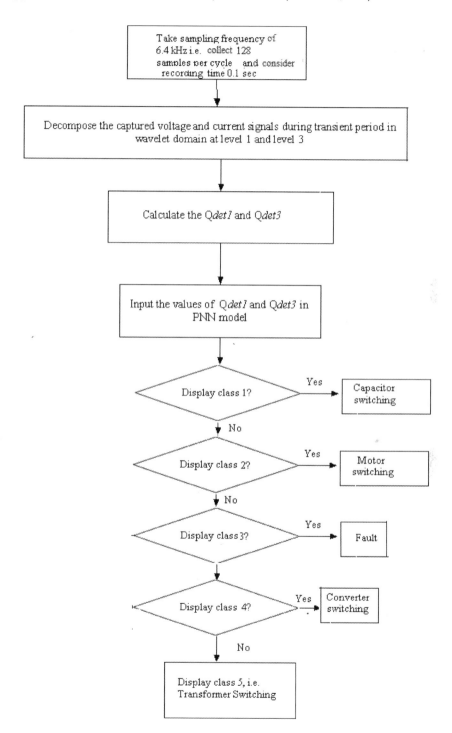

FIGURE 10.1 Flowchart for classification of different transients.

for categorization of different transients. A PNN is chosen as a classifier to distinguish the transient signals. Every input signal is created with 128 samples per cycle; 0.1 second is the capturing time and consequently 6.4 kHz the sampling frequency.

DWT is used for calculation of reactive power details at the sending end at levels 1 (Q_{det1}) and 3 (Q_{det3}) using the equation,

$$Q_{det} = \frac{1}{2^p} \sum_{k=0}^{2^{j0}-1} d_{j0,k} d``j0,k$$

These reactive powers are utilized as the input to the PNN model for classification of transients. In this method, motor switching, capacitor switching, fault, converter and transformer switching are considered classes 1, 2, 3, 4 and 5, respectively, during training of the PNN. So according to the input, this algorithm will display the corresponding class of the transient signal. The reliability and validity of numerous case studies have been tested, which indicates that this method is ideal for classifying transients.

10.3 SYSTEM DESCRIPTION

In Figure 10.2, a radial distribution network is considered. For each section Z_S, the internal impedance of the voltage source and the line impedance are taken as (0.5 + j0.5) Ω and (1 + j1) Ω, respectively. Each load impedance, that is, Z_1, Z_2, Z_3, Z_4 and Z_5 is (300 + j300) Ω. The transformer, capacitor, motor and converter are joined to time-controlled switches to every bus to get detailed reactive power under diverse transient conditions. Electromagnetic transient program (EMTP) is implemented to get the instantaneous current and voltage signals from the sending end, which is shown in Figure 10.2. The signals are analyzed using DWT with MATLAB.

11 kV distribution systems are modeled. Figure 10.2 shows the EMTP model. The capturing time is 0.1 second, and 6.4 kHz is the sampling frequency, that is, 128 samples per cycle. A Daubechies wavelet with four filter coefficients (db4) is used for feature extraction. Simulation of all five transients is performed. The switching transients are classified by the PNN. The original signal is decayed up to level 5 with discrete wavelet transform, and then after consideration of all the parameters, the detailed reactive power at levels 1 and 3, that is, Q_{det1} and Q_{det3}, are chosen for the classifier as inputs. These are calculated using the equation,

$$Q_{det} = \frac{1}{2^p} \sum_{k=0}^{2^{j0}-1} d_{j0,k} d``_{j0,k}.$$

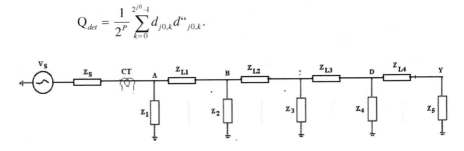

FIGURE 10.2 Radial network for classification of power quality events.

10.4 WAVELET TRANSFORM-BASED FEATURE EXTRACTION

10.4.1 CAPACITOR SWITCHING

Figure 10.3 shows the capacitor-switching transients in the network of Figure 10.2. Capacitors of 50 to 550 μF are considered during the simulation of capacitor-switching transients. Various Q_{det} values from levels 1 to 5 at various buses are shown in Table 10.1 after switching on the capacitor at 150 μF. Current and voltage signals

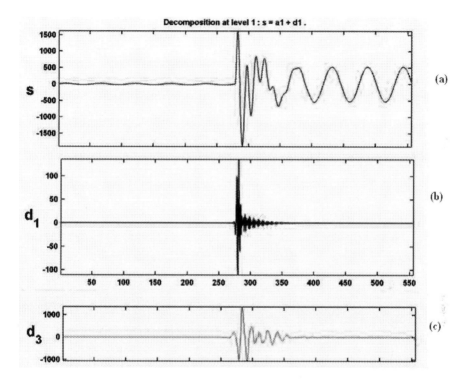

FIGURE 10.3 (a) Transient in current signal for capacitor switching, (b) signal decomposed at level 1, (c) signal decomposed at level 3.

TABLE 10.1
Q_{det}**Values for Capacitor Switching at Different Buses**

BUS	Q_{det1}	Q_{det2}	Q_{det3}	Q_{det4}	Q_{det5}	Class
A	1.64×10^8	1.67×10^6	3.87×10^7	2.2×10^6	4.3×10^7	1
B	1.07×10^7	1.87×10^6	2.78×10^7	2.4×10^6	$3.24.3 \times 10^7$	(C=150 μF
C	4.07×10^6	1.9×10^6	2.78×10^7	2.01×10^6	$3.34.3 \times 10^7$	Switching angle 60°)
D	1.98×10^6	2.1×10^6	2.78×10^7	1.97×10^6	$3.34.3 \times 10^7$	
Y	1.05×10^6	2.44×10^6	2.78×10^7	2.3×10^6	$3.34.3 \times 10^7$	

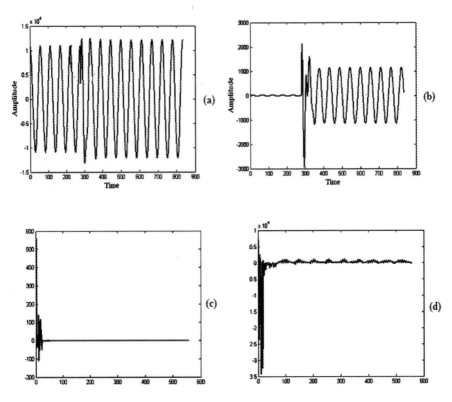

FIGURE 10.4 (a) Voltage transient during capacitor switching, (b) current during capacitor switching transient, (c) detailed reactive power at level 1, (d) detailed reactive power at level 3.

are taken at only the sending end. The voltage profile and current during capacitor-switching transients are shown in Figure 10.4(a) and (b), respectively. These voltage and current signals are decomposed up to level 5 utilizing DWT; then, according to the equation, $Q_{det} = \frac{1}{2^P} \sum_{k=0}^{2^{j0}-1} d_{j0,k} d''_{j0,k}$, harmonic reactive powers are calculated in each decomposition level. In Figure 10.4 (c) and (d), the detailed harmonic reactive powers are shown at levels 1 and 3.

10.4.2 Motor Switching

Figure 10.5 shows the motor-switching transients. To perform the simulation, induction motors with ratings 5 to 10 kW are taken, and the switching time is also changed for collection of a sufficient amount of data.

Various Q_{det} values from levels 1 to 5 at various buses are shown in Table 10.2 after switching on the 9-kW induction motor at its peak voltage. In Figure 10.5(a) and (b), the voltage and current profile during motor switching transients are shown. In

Application of PNN and Wavelet Analysis to Classify Power Quality 223

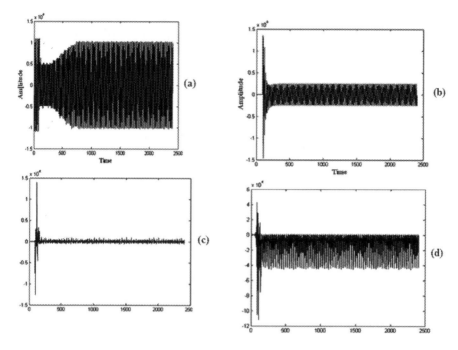

FIGURE 10.5 (a) Voltage profile during motor switching transient, (b) current during motor switching transient, (c) detailed reactive power at level 1, (d) detailed reactive power profile at level 3.

TABLE 10.2 Q_{det} Values for Motor Switching at Different Buses

BUS	Q_{det1}	Q_{det2}	Q_{det3}	Q_{det4}	Q_{det5}	Class
A	2.1×10^8	6.1×10^8	2.9×10^7	5.5×10^8	2.4×10^7	2
B	6.1×10^7	4.21×10^8	6.76×10^6	3.75×10^8	6.2×10^6	(9 kw
C	4.6×10^7	3.3×10^8	4.05×10^6	2.56×10^8	3.67×10^6	Switching angle 90°)
D	3.3×10^7	2.4×10^8	3.45×10^6	1.87×10^8	2.78×10^6	
Y	2.7×10^7	9.7×10^7	2.56×10^6	9.1×10^7	1.97×10^6	

Figure 10.5(c) and (d), detailed reactive power profiles at levels 1 and 3 are shown for motor-switching transients.

10.4.3 FAULT-INDUCED EVENTS

Figure 10.6 shows the fault-induced transients and various Q_{det} values from levels 1 to 5 when asymmetrical faults occur at various buses, shown in Table 10.3. In Figure 10.6(a) and (b), the voltage and current during fault-induced transients are shown. In Figure 10.5(c) and (d), detailed reactive power at levels 1 and 3 are shown during fault-induced transients.

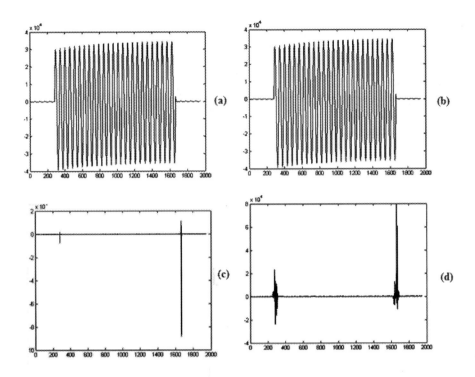

FIGURE 10.6 (a) Voltage transient during fault induced transient, (b) current profile during fault induced transient, (c) detailed reactive power profile at level 1, (d) detailed reactive power profile at level 3.

TABLE 10.3
Q_{det} Values for Fault-Induced Transients at Different Buses

BUS	Q_{det1}	Q_{det2}	Q_{det3}	Q_{det4}	Q_{det5}	Class
A	2.5×10^5	1.2×10^6	2.3×10^6	1.1×10^6	6.8×10^5	3
B	1.1×10^5	1.2×10^6	7.2×10^6	4.2×10^5	3.2×10^5	($R_f = 20\ \Omega$ Switching angle 90°)
C	4.2×10^4	5.67×10^5	4.2×10^6	2.1×10^5	2.1×10^5	
D	3.98×10^4	2.95×10^5	3.1×10^6	1.1×10^5	1.06×10^5	
Y	2.44×10^4	1.48×10^5	2.1×10^6	6.7×10^4	2.08×10^4	

10.4.4 Converter Switching

Transients produced through converters are shown in Figure 10.6, as well as 12 pulse converters with changing triggering angles. Various Q_{det} values from levels 1 to 5 at various buses are shown in Table 10.4.

In Figure 10.7(a) and (b), the voltage and current profile during converter switching transients are shown, and in Figure 10.7(c) and (d), detailed reactive power profiles at levels 1 and 3 are shown during converter-switching transients.

Application of PNN and Wavelet Analysis to Classify Power Quality 225

TABLE 10.4 Q_{det} Values for Converter Switching at Different Buses

BUS	Q_{det1}	Q_{det2}	Q_{det3}	Q_{det4}	Q_{det5}	Class
A	5.21×10^3	4.9×10^4	1.68×10^6	4.49×10^4	5.45×10^7	4
B	3.61×10^3	4.77×10^4	1.68×10^6	4.51×10^4	4.56×10^7	(α = 60°
C	1.53×10^3	4.58×10^4	1.68×10^6	4.74×10^4	3.45×10^7	firing angle)
D	1.25×10^3	4.55×10^4	1.68×10^6	4.77×10^4	2.34×10^7	
Y	1.2×10^3	4.54×10^4	1.69×10^6	4.61×10^4	1.12×10^7	

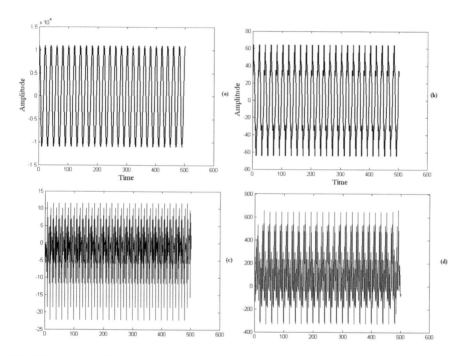

FIGURE 10.7 (a) Voltage profile during converter switching, (b) current during converter-switching transient, (c) detailed reactive power at level 1, (d) detailed reactive power at level 3.

10.4.5 Transformer Switching

Transformer-generated transients are shown in Figure 10.8. Transformers with various kilo-volt-ampere(KVA)ratings are selected to differentiate the inrush of transformer from other transients. Various Q_{det} values from levels 1 to 5 are shown in Table 10.5 after switching on the 100-kVA transformer. In Figure 10.8(a) and (b), the voltage and current profile during transformer-switching transients are shown. In Figure 10.8(c) and (d), detailed reactive power profiles at levels 1 and 3 are shown during transformer-switching transients.

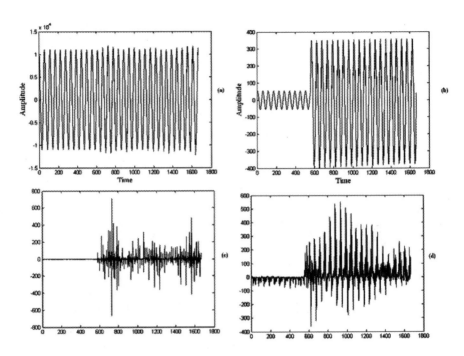

FIGURE 10.8 (a) Voltage transient during transformer switching, (b) current during transformer switching transient, (c) detailed reactive power at level 1, (d) detailed reactive power at level 3.

TABLE 10.5
Q_{det} Values for Transformer Switching at Different Buses

BUS	Q_{det1}	Q_{det2}	Q_{det3}	Q_{det4}	Q_{det5}	Class
A	1.05×10^5	5.98×10^5	1.83×10^7	6.07×10^7	6.2×10^5	5
B	1.04×10^5	5.97×10^5	1.78×10^7	5.99×10^7	6.01×10^5	(10 KVA
C	8.9×10^4	5.85×10^5	1.78×10^7	5.2×10^7	5.9×10^5	Switching
D	1.09×10^5	5.2×10^5	1.79×10^7	3.45×10^7	5.28×10^5	angle 60°)
Y	1.17×10^5	5.15×10^5	1.78×10^7	2.3×10^7	5.6×10^5	

10.5 CATEGORIZATION OF TRANSIENTS USING PNN

Simulation is carried out to classify the transients. Each type is considered as an individual class and assigned a classification number from 1 to 5. To calculate the DWT reactive power, MATLAB code is implemented, that is, Q_{det1}, Q_{det2}, Q_{det3}, Q_{det4} and Q_{det5}. Reactive power details are used as input to the PNN model. Seventy-five feature sets are obtained for each transient, from which 25 sets are utilized for training purposes such that 125 training samples are taken along with another 50 sets so 250 features are utilized for testing purposes. For every transient, to find testing and training data for the PNN model, the authors changed the capacitor capacitance value, induction motor kW rating, converter firing angle, transformer kVA rating, fault resistance and fault occurrence time.

Simulations gave 375 sets of detailed reactive power (Q_{det}) up to level 5 individually for the previous five transients. The specified groups of transient samples are given in Table 10.6. In conclusion, the rate of success for classifying the various types of transients also depends on the degree of decomposition. The success rate is characterized by adjusting the decomposition level in Table 10.7. This table indicates that the success rate remains the same whether the original signal is broken to level 5 rather than level 3. In the first, second, third and sixth rows, the success rate is also considered similar. However, the success rate is relatively low in the remaining rows. So the authors selected levels 1 and 3 to lower the computational load in view of all the findings.

TABLE 10.6
Classes of Transient Samples and Simulated Data Set for Training and Testing of PNN Network

Type of Transient	Assigned Class	Training Data (Set)	Testing Data (Set)
Capacitor switching	1	25	50
Motor switching	2	25	50
Fault	3	25	50
Converter	4	25	50
Transformer switching	5	25	50
Total		125	250

TABLE 10.7
Decomposition Level and Success Rate

SL. No	Decomposition Level	Features Selected	Success Rate (%)
1	Decomposed up to level 5	Q_{det1} to Q_{det5}	99.6
2	Decomposed up to level 4	Q_{det1} to Q_{det4}	99.6
3	Decomposed up to level 3	Q_{det1} to Q_{det3}	99.6
4	Decomposed up to level 2	Q_{det1} and Q_{det2}	92.1
5	Decomposed at level 1	Q_{det1} only	21.4
6	Decomposed at levels 1 and 3	Q_{det1} and Q_{det3} only	99.6
7	Decomposed at levels 1 and 2	Q_{det1} and Q_{det2} only	97.6

TABLE 10.8
Comparative Study with Other Existing Methods

Sl. No.	Method for Identification of Power System Transients	Success Rate
1	Wavelet-transform+ANN based classifier[33]	95
2	Wavelet-transform based classifier [34]	95
3	S-transform +competitive neural network based classifier [35]	97.5
4	DTCWT+ ANN based classifier [6]	100
5	Proposed method	99.6

TABLE 10.9
Confusion Matrix

Class	1	2	3	4	5
1	50	–	–	–	–
2	–	49	1	–	–
3	–	–	50	–	–
4	–	–	–	50	–
5	–	–	–	–	50

The results of the classification can be represented by a confusion matrix, which is standard for classifying studies. Table 10.8 indicates that there are one row and one column in an uncertainty matrix for every class. The row indicates the original class and the column reflects the PNN classification predicted class. In the matrix, the number displays the classification of different patterns observed from the test sample. For example, in Table 10.8, only 1 of 50 cases of motor switching was classed as faulty. For every transient, the error rate is defined in the table of the network classification. It can be concluded from these findings that the network properly classifies 249 of the 250 samples using this approach. This means that 99.6% is the correct classification rate. A plurality of grouping errors exists between classes 2 and 3 in Table 10.8, since they are identical in terms of characteristic harmonics.

10.6 COMPARATIVE ANALYSIS

Table 10.9 demonstrates the comparative study with other existing methods. These methods are to some extent like the method proposed by the authors in the sense that the features of distorted signals were extracted using WT and then classified using ANN.

The results presented in Table 10.9 show the proposed approach as well as better performance with a better success rate, which is 99.6%, along with lower computational complexity.

10.7 CONCLUSIONS

In this chapter, an original, easy approach is proposed based on PNN and DWT for categorization of different switching transients. A precise and consistent power transient classifier is established to classify such transient signals. DWT was used to calculate the level 1 (Q_{det1}) and 3 (Q_{det3}) comprehensive reactive power utilized as the ANN model input. A probabilistic neural network is utilized as a classifier. Its reliability as well as the validity of numerous case studies is tested and indicates that this method is ideal for classifying the transients in power system networks.

REFERENCES

1. Sarıbulut, L. A simple power factor calculation for electrical power system. *Electric Power Energy System*, 62, pp. 66–71, 2014.
2. Zwe-Lee, Gaing. Wavelet-based neural network for power disturbance recognition and classification. *IEEE Transactions on Power Delivery*, 19(4), pp. 1560–1568, 2004.
3. Beg, M. A., Khedkar, M. K, Paraskar, S. R., and Dhole, G. M. Feed-forward artificial neural network–discrete wavelet transform approach to classify power system transients. *Electric Power Components and Systems*, 41, pp. 586–604, 2013.
4. Mao, P. L., and Agarwal, R. K. A novel approach to the classification of the transient phenomena in power transformers using combined wavelet transform and neural network. *IEEE Transactions on Power Delivery*,16, pp. 6654–660, 2001.
5. Chakraborty, S., Chatterjee, A., and Goswami, S. K. A sparse representation based approach for recognition of power system transients. *Engineering Applications of Artificial Intelligence*, 30, pp. 137–144, 2014.
6. Chakraborty, S., Chatterjee, A., and Goswami, S. K. A dual tree complex wavelet transform based approach for recognition of power system transients. *Expert Systems*. http://dx.doi.org/10.1111/exsy.12066.
7. Mokryani, G., Siano, P., and Piccolo, A. Detection of inrush current using S-transform and competitive neural network. In: Proceedings of the 12th International Conference on Optimization of Electrical and Electronic Equipment, 2010.
8. Ibrahim, W. R. A., and Morcos, M. Artificial intelligence and advanced mathematical tools for power quality applications: A survey. *IEEE Transactions on Power Delivery*, 17, pp. 668–673, April 2002.
9. Kucuktezan, C. F., and Genc, V. M. I. Dynamic security assessment of a power system based on probabilistic neural networks. In: Innovative Smart Grid Technologies Conference Europe (ISGT Europe) IEEE PES, Gothenburg, 11–13 October 2010.

11 Overview of Security and Protection Techniques for Microgrids

*Asik Rahaman Jamader, Puja Das,
Biswaranjan Acharya, and Yu-Chen Hu*

CONTENTS

11.1	Introduction	232
11.2	Background	234
	11.2.1 Key Characteristics of a Microgrid	234
	11.2.2 Topology of Microgrids	234
	11.2.3 Types of Microgrids	235
	11.2.4 DER Elements	237
	11.2.5 Control of Microgrids	237
11.3	Structure of Microgrids	238
11.4	Problems in Microgrid Safety Systems	239
	11.4.1 Shifting Aspects in Error Importance	240
	11.4.2 Loss of Mains	240
	11.4.3 Avoidable Discontinuations	240
	11.4.4 Blinding of Security	240
	11.4.5 Changes in the Short-Circuit Level	242
	11.4.6 Prohibition of Programmed Reclosing	243
	11.4.7 Unsynchronized Reclosing	243
	11.4.8 Wrong Stumbling	243
11.5	Different Methods of Protecting Microgrids	243
	11.5.1 Distance Protection	244
	11.5.2 Multi-Agent	246
	11.5.3 Current Regulator	247
	11.5.4 Variables Used in Protection	248
	11.5.5 Adaptive Protection	248
	11.5.6 Centralized Defense Scheme	249
	11.5.7 Decentralized Defense Scheme	249
11.6	Discussion	250
11.7	Future of Microgrids	250
11.8	Conclusion	250
References		252

DOI: 10.1201/9781003121626-11

11.1 INTRODUCTION

Today, interest in electrical energy is persistently developing. To fulfill the growing need, more power plants must be developed. Normal energy or power plants have downsides, for example, carbon emissions, low proficiency, high transmission misfortunes, higher development and fuel cost, high development time and lower liability [1]. Consequently, the idea of a microgrid was developed, interfacing a few smallergenerators called miniaturized scale sources/distributed generators (500kW) overcharges turbines, sun-based force plants, wind energy units, smaller-scale turbines, small hydroelectric force plants, collective heat plus power as clusters, close to the low-voltage (LV) side to meet the necessary requirements.

A microgrid is an energy-spreading complex network including distributed generation assets, various loads at the voltage level of appropriation and energy loading components. From a system point of view, a microgrid is beneficial in light of the fact that it is a controlled unit (CU), even though it can be misused as concurrence-evaluated capacity. From the perspective of clients, a microgrid may be intended to address uncommon necessities, for example, unwavering higher quality, better neighborhood voltages, expanded productivity, voltage hang rectification and unit-perceptible influence flexibility. From an ecological point of view, a microgrid reduces natural contamination and the Earth-wide temperature boost since it creates less carbon monoxide (CO) [2, 3]. Regardless of how a microgrid is a fitting sub statute for constrained non-renewable energy sources and can proficiently take care of intensity age issues, it is still generally confined to a research facility scale because of various technical difficulties [4].

The most generally significant of these difficulties are protection, security, power excellence, activity in typical and islanded modes, voltage and recurrence regulators, attachment and play activity, vitality of the executive functions and framework solidity [5, 6, 7]. Table 10.1 lists a comparative analysis between the current grid and smart grids.

TABLE 11.1
Comparative Analysis between Current Grid and Smart Grids

SL NO.	Current Grid System	New Intelligent Grid System
1	Electromechanical	Digital
2	Integrated Genesis	Distributed Genesis
3	Limited Sensors Device	AL Overall Sensors
4	One-Way Traffic	Two-Way Traffic
5	Hierarchical	Network
6	Blind	Self-Monitoring
7	Few Client Adoptions	Numerous Client Adoptions
8	Partial Control	Universal Control
9	Fails and Blackouts	Adaptive and Islanding
10	Physical Assessment	Distant Assessment

Security and Protection Techniques for Microgrids 233

An adoption framework that works as a microgrid brings an adequate age near the load and thus can keep up the force gracefully in the case of macrogrid issues. Likewise, if a microgrid has an overabundantage limit, it can give the macrogrid framework recovery assets, in this way shrinking the recurrence and span of blackouts. Such highlights add adaptability to the previously mentioned engineering and give microgrids a huge advantage for improving network flexibility to macrogrid failures. In any case, the advantages presented by a microgrid will be in danger if the microgrid isn't appropriately ensured during the short-out errors that happen inside its own limits. In spite of the fact that microgrids are raised at the circulation level of the inheritance matrix, accessible delivery defensive equipment, for example, meld and reclose [4], are not an appropriate possibility for microgrid security because of different reasons, a significant part of which is very truthful and corporate defensive along with a variety of strategies have been distributed in the present form. The lower limit of the macrogrid can be bigger in some cases of the microgrid through more than one request for size, making the flaw current levels over the macro grid-associated and islanded methods of a microgrid pointedly extraordinary. Such a major dissimilarity makes organization of the current dissemination of defensive equipment troublesome and regularly out of reach.

A variety of strategies have been distributed over the most recent couple of years to address microgrid promise issues [7, 8, 9]. In the interim, given the affectability of dependable assurance, broad testing and checks are required for a security plan to arrive at a business level. Even distributed exploration on microgrid protection has not approached this level yet. Without readymade microgrid transfer, defensive tools of sub-transmission and transmission structures, for example, directional overcurrent, departure and variance transmissions, are the main decisions that can make a protection framework progressively invulnerable to the difficulties of microgrids, for example, factor cut-off level, but at a greater expense. So, In spite of that there must be have some construct between all attribute of microgrid. These transfers are required to be the primary alternatives accessible to secure microgrids in any event for the following years. Figure 11.1 depicts the characteristic layer arrangement of the microgrid.

These transfers are supported by many years of effective activity for microgrids; thus, the correct activity for microgrids is regularly underrated [10]. In any case, a key factor in the improvement of the vast majority of these transfers is that the age units incorporate straightforwardly coupled synchronous machines (SMs), a commonly unacceptable idea for the DER components of a microgrid. Accordingly, the impacts of DER parts, such as specific special qualities of microgrids on readymade defensive transfers, should be reviewed to recognize the fundamental services and adjustments to these transfers. These are fundamental steps to guarantee compact security of microgrids and thus the framework strength.

This chapter is focused on the request of in-demand accessibility for microgrid security. Section 11.2 gives the background of this chapter. Section 11.3 discusses the structure of the microgrid. Section 11.4 explains challenges in microgrid protection systems, which consist of different challenges to protect microgrids. Section 11.5 contains a method to protect microgrids. Lastly, Sections 11.6 and 11.7 contain discussion and the conclusion.

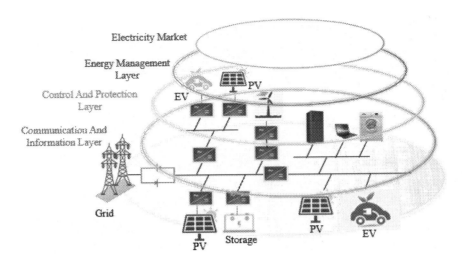

FIGURE 11.1 A characteristic layer arrangement of the microgrid.

11.2 BACKGROUND

This section surveys various classes and methods of microgrids and their segments. Figure 11.2 demonstrates the graphical chart of a microgrid that incorporates double equal feeders radiating from the point of common coupling on the microgrid.

11.2.1 Key Characteristics of a Microgrid

- Connecting to the normal framework is *optional*
- Flexible, steady quality and supportability are the center's responsibilities.
- Backup for all framework loads, not simply basic loads.

Current innovations are expected to enhance vitality creation and use.

DER units and loads inside the microgrid are equipped with nearby controllers. The level of improvement and functionalities of supervisory controller and energy management system (SC-EMS) can change broadly. As of now, there are no rules to smooth out such capacities, which can have huge consequences for the protection of microgrids.

11.2.2 Topology of Microgrids

The microgrid structure can be circular, assembled or blended [11]. Topology is an element that influences the title and scale of issuing current and insurance methods in the microgrid. For example, a defective current separates in two equal ways but in a circular manner. Also, an upstream feeder detects double current on protection tools in every route inside a circle. On the other side, the current flow in the mess topology is equivalent for both upstream and downstream branches [12]. This chapter proposes protective measures for DC microgrids with circle topology.

Security and Protection Techniques for Microgrids

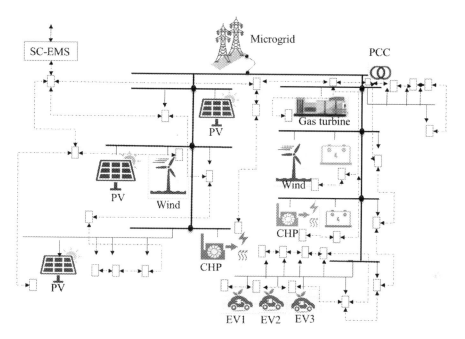

FIGURE 11.2 Diagram of a microgrid with connection.

The topology of the microgrid structure is shown in Figure 11.3. Its essential units include a circulation system, a bidirectional inverter, a vitality stockpiling part, a PV and fan power age and load, a sum of the five basic principles of the unit.

11.2.3 Types of Microgrids

Considering the qualities/features of the feeders which form the foundation of a grid system, they are classified into three main points given in the following.

Urban Microgrids: Here, feeders are located in compact mechanized areas or settled; these are heavily loaded, the primary trunk and side parts are adequately compacted and the grade of inequality is not high. The ratio of short-circuit in this case, with its PCC, is usually greater than 25. In this method, the voltage and fluctuation of a microgrid during microgrid-related processes is controlled by the grid system, and the profile of voltage is actually smooth.

In this way, in some cases, synchronization is noticed by the electronically coupled distributed generation (ECDG) part in the microgrid, which is important for reliability.

Rural Microgrid: Insufficiently populated areas are the location of feeders in this class, so the load is dismantled. The short-circuit ratio is not really higher than that of the urban microgrid, and the profile of the voltage is not smooth. In this microgrid, abnormalities of voltages and changes can be noticeable. Therefore, DER elements

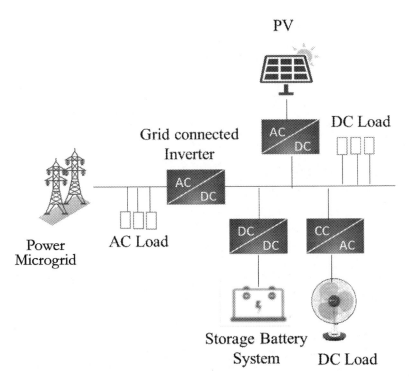

FIGURE 11.3 Microgrid topology.

have impact the power (volts), and whenever allowed, the feeder voltage can be controlled per guidelines.

Off-Grid Microgrids: Off-network microgrids are located in geographically remote areas where there is no scope for microgrid links or which are bounded by problematic scenarios for broadcast route associations. To be specific, the off-grid microgrid functions continuously in islanded mode and so does not conform to the exact meaning of "microgrid." It actually does not obey the rules of microgrid-associated activity. The incorporation of large DER elements with off-grid microgrids is happening faster than urban and provincial microgrids, as is work on SC-EMS procedures for off-network microgrids. There is serious social, financial, political and natural impulses and tension concerning the coordination of sustainable power sources with urban and country microgrids and off-grid microgrids, which are being built entirely for islanded mode. Thus, this type of microgrid will primarily be the fastest growing in the microgrid family, considering all the difficulties in the microgrid family.

Most contracted scale networks can be one of five classes (as characterized by the Microgrid Institute):

- Off-grid microgrids that are islanded and have remote destinations and small-scale medium frame brushes are not associated with neighboring utility systems.

- Campus microgrids that are fully connected to a neighborhood utility mesh but can keep some degree of administration in isolation from the network, for example, during a utility blackout. Mill models run college and corporate grounds and containment facilities and serve military installations.
- Community microgrids that are involved in utility systems. Such microgrids serve different clients or administrations within a network and provide robust capacity to largely unavoidable network resources.
- District energy microgrids that give power just as heat strength for warming (and cooling) several offices.
- Nanogrids incorporate the smallest isolated structure units with the skill to work liberally. A Nanopattern is characterized as an introverted configuration or a private strength gap.

11.2.4 DER Elements

DER elements are electronically paired and divided into spinning machine units. The main purpose of this classification is that the interface medium shows as much reliability and control as the conventional qualities of a unit.

ECDG Units: In this type, an alternative or direct current, a voltage sourced generator, mediates between the microgrid and origin [11, 12]. A voltage-sourced converter is divided into either a single stage or three stages depending on the method of energy creation. Type-4 wind turbine frameworks, voltage-sourced converter PV frameworks, class-4 wind turbine frameworks and medium-sized rapid gas-turbine-machine frameworks use distributed generation techniques. This can be non-dispatch able or dispatch able units. MPPT controls the supply for a non-dispatch able unit [13].

Pivoting Machine-Coupled Distributed Generation Sections: Spinning machines are adopted and incorporated by this section with the following points:

- Class-1 SCIM utilized by WTG
- Class-3 DFIM merged with WTG; diesel-producer elements, which receive recorded, controlled SM

11.2.5 Control of Microgrids

The control of microgrids includes three types of method: centralized, scattered and progressive controls [14]. Certain unmistakable properties of the microgrid, for example, the maximum level of irregularity and the mixed variety of DER elements, emphasize the difficulties in formulating an appropriate control for each working condition.

In addition, two special task modes (i.e., islanded and connected modes), plus the versatile nature of microgrid control, can be further enhanced as needed. Meanwhile, the regulation of automatically coupled elements has a great influence on the grid's temporary conduct, especially due to shortcomings, which inevitably disturb the protective parts of the grid system. A basic step taken to handle microgrid regulation/security issues is to distinguish different methods of activity, as described in the following.

Macrogrid-Connected Approach: In this mode, every DER element incorporates a load controller that relies on the component's privately predicted current and voltage and their consequences. SC-EMS of microgrids produce DER orientation waveforms that are expected to organize the activity of DER elements. In this mode, SC-EMS is similarly a choice to obtain data from the macrogrid.

Islanded Approach: Here, in operation, for internal activity only, LC and SC-EMS are used in the microgrid. The regulator methods of microgrid-related islands and macrogrid-related systems are often remarkable, as in macrogrid-related modes, all DER elements can function in PQ-switch method. Control progress should be addressed and monitored through SC-EMS.

Mesogrid Approach: A mesogrid is characterized as a bunch of microgrids, each with an SC-EMSs and LSs, which are associated with a host macrogrid and work in a pre-specified facilitated way. Each microgrid inside the mesogrid works as a simulated force unit dependent on certain orders that are created through its SC-EMS and imparted to LCs inside the microgrid. The upper layer of the SC-EMS, the mesogrid SC-EMS, gets data from the macrogrid and from every individual microgrid and, on that premise, determines the control/activity duties of each microgrid concerning its PCC on this basis.

11.3 STRUCTURE OF MICROGRIDS

A microgrid comprises three important parts: miniaturized scale generators (according to the paradigm, a PV cluster and power module producer), neighborhood storage components and various loads. A microgrid can be a private or three-stage framework; it might be associated with minimal or mid-voltage dissemination organization and be able to work in regular and islanded modes [15]. A graphical portrayal of an example microgrid is shown in Figure 11.4 [16]. Figure 11.4 defines the DC, AC, and high-frequency AC atmosphere.

ADC microgrid is a movement structure linking DC loads, energy storage components and distributed generation assets that are normally bottomless, with DC voltage as outcome. On the other hand, an AC microgrid comprises AC transports, is suitable for mass association and loads to its belongings. These transports are used as bidirectional converters with the help of the DC method. Figure 11.4 defines the zonal position as 1 for DC and zones 2 and 3 for AC microgrids. DC microgrids utilize a convertor in both DC and AC classifications and as a crossing point from DC to AC microgrids [17]. The associated distributed sustainable power source belonging to DC transports does not moderate the universal production of the organism, because a reduction in the extent of converter rub-down lowers the requirement of vitality altering, but it expands the strained nature of the arrangement.

High-frequency alternative current (HFAC) frameworks are now used in aircraft and train technology. The use of frequencies from 400 Hz to 20 kHz for power allocation frameworks is currently under investigation. The preferred position of HFAC frameworks is that the size is very small, with extraordinary power depth, and private. However, the application of these frameworks to microgrids is uncertain because maximum repetition can result in maximum power misalignment and a voltage drop with the current.

Security and Protection Techniques for Microgrids

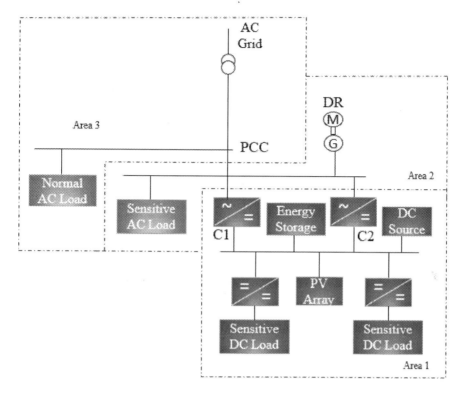

FIGURE 11.4 Sample of AC and DC microgrid.

11.4 PROBLEMS IN MICROGRID SAFETY SYSTEMS

To get better protection, a selection will be made for strong security activities, theoretical activities, stabilization, adaptation, legalization of rational equipment related to the best rankings and less testing. As surveyed before, conventional protection plans are intended for redial streams with high shortcomings. In reference to the previous discussion, the conventional protection schemes outlined for radial flow with high fault currents for distribution networks will not function without error in a microgrid because of bilateral power flow, evolving characteristics of distributed generations, sporadic nature of distributed generations and variation in fault current [18]. The crucial challenges faced in the protection of microgrids are:

1. Shifting aspects in error importance
2. Loss of mains
3. Avoidable discontinuations
4. Blinding of protection
5. Changes in the short-circuit level

6. Exclusion of routine reclosing
7. Unsynchronized reclosing
8. Prohibition of automatic reclosing

11.4.1 Shifting Aspects in Error Importance

The distributed generation connection in LV has been modified to a significant level of the main level because of privacy and islanded modes. In grid-combined mode, the event is much higher because both the internal use of the microgrid and the distributed generation signal are available. In any case, there is a liability in the slow speed of time, and the boundaries of the original source in the microgrid are determined.

The current fluctuation depends on what distributed generation type is used [19]. Additionally, new natural source-based distributed generations exist as exceptionally long-term distributed generations, as a result these distributed generations are simply imputing the current situation. Therefore, the error magnitude of the current depends on the method of operation. As a result, it is difficult to accurately estimate the current state of the error.

11.4.2 Loss of Mains

Loss of mains (LOM) defines the discontinuation utility of microgrid activation from the origin, although it is associated with the part of the utilized gate. (1) Error of utility grid and (2) issue in circuit roller associating with the main utility. During such unplanned islanding, since a piece of the system is charged and the islanding sun identified, there is a danger of error [18]. Additionally, this procedure brings about uncontrolled voltage and recurrence, dangers to plants accessible in the islanded framework and unsynchronized terminations which eventually bring in the destruction of client apparatus and computing devices.

11.4.3 Avoidable Discontinuations

This is significant when a distributed generation is found near a substation and in any occasion when there is a crisis near the substation of the neighboring feeder (Figure 11.5); at that point, in such a case, significant portion of the deficient current is contributed by the distributed generation.

This makes the current surpass its pickup estimation accessible in the solid feeder (Relay1), and subsequently it stumbles before the hand-off compared to the faulty feeder (Relay2).

11.4.4 Blinding of Security

Both the utility and the distributed generation connect the critical current when there is a problem with the bottom completion of the feeder in a microgrid. Currently, Thevenin's barrier given at the error point has been extended in contrast to the foundation method due to the extra barrier given through the distributed generation.

Security and Protection Techniques for Microgrids

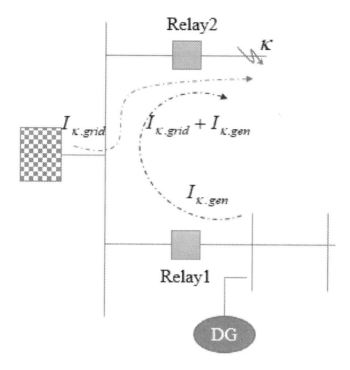

FIGURE 11.5 Avoidable discontinuations.

In Figure 11.6, a small distributed generation is added some way off. For a lack located in a well far from L, each stage is determined as the most extreme issue.

$$I_k = \frac{Vot_{th}}{\sqrt{3}Z_{th}} \tag{11.1}$$

where Vot_{th} is the voltage fault at the faulty point, and Z_{th} is Thevenin's low impedance. Let z_o, z_g, z_L be different states defining origin of utility, impedance of distributed generation and transmission line, respectively. At that point, Thevenin's proportional circuit of the system can be shown as Figure 11.7. Thevenin's low is determined as

$$Z_{th} = \frac{(Z_s + l.Z_L).Z_g}{(Z_s + l.Z_L + Z_g)} + (1-l)Z_L \tag{11.2}$$

Therefore, Thevenin's impedance offered at the blamed point is expanded because of the extra impedance offered by distributed generation. The current contributed from the lattice is

$$I_{grid} = \frac{Z_g}{(Z_s + l.Z_L + Z_g)}.I_k \tag{11.3}$$

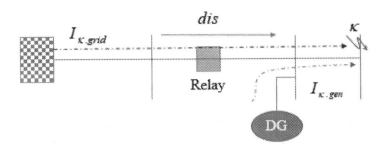

FIGURE 11.6 Blinding of assurance.

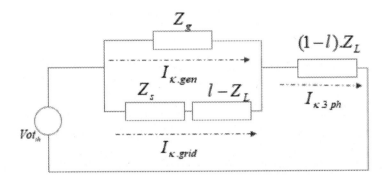

FIGURE 11.7 Thevenin's equivalent circuit.

The promise of the current deficit by the usefulness is not linear with the field and size of the distributed generation. Later, when an error arises at the bottom of the feeder in a microgrid, the lattice barrier will be as wide as the distributed generation barrier, and so the short-circuit current below the feeder hand-off surprise (pick) current in the LV is good, which causes the connection to identify the problem and thus the entire insurance structure. So in order to overcome the issues described previously, it is necessary to create a valid insurance system to withstand the problems described previously and to work efficiently. As a result, the region audits various security plans accessible to the microgrid.

11.4.5 Changes in the Short-Circuit Level

The ability to operate the microgrid in protected and islanded mode is an important test for the appropriate security plan. This problem causes significant differences between constraints in these modes. In normal operation mode, a microgrid is connected to a medium voltage (MV) array and, to provide the resulting flow, both the system and the distributed generation are connected. Due to the need for circuit breaker (CB) replacement, the commitment of the distributed generations may delay the system above its approved respect. In islanded group mode, the microgrid separates from the MV system. Thus, in small events, only the distributed generation lacks flexibility. Since the estimate of the problem provided by the director general is limited [21], the current level of error in this landslide mode is fundamentally lower than in normal mode. This level may be peculiar to traditional insurance structures, which are designed based on the size of the current deficit and may not be able to identify problems. This problem requires a strategy that can ensure microgrids in both normal and geo-mode.

11.4.6 PROHIBITION OF PROGRAMMED RECLOSING

The scattered system spreads without the distributed generation. Then, when restoration work is done, the flowing part of it is detached to clear transient errors. In the presence of the distributed generation, both the MV system and the distributed generation are strongly deficient. However, the recovery separates the MV system, the distributed generation handles the current deficit until the error is recovered and, in this situation, the transient problem turns into a permanent issue.

11.4.7 UNSYNCHRONIZED RECLOSING

At this point, when a distributed generation is connected to a system, two unknown structures are closely linked. If the association is run without thinking about synchronization, it can damage the micro-hardware by extending the distributed generation. It is important to note that the "single wire earth return" promotions are seen as nothing but "waiting issues" that empower easily accessible customers, only through direct access or receiving via earth [22].

11.4.8 WRONG STUMBLING

Bogus stumbling happens when the shortcoming current happening in a feeder is enhanced by the flawed current produced through a distributed generation in a neighboring feeder appended to a similar sub-station. In this situation, the defensive hardware of the neighboring feeder may disengage the circuit, producing an issue called pointless blackout of feeder or bogus stumbling [23].

11.5 DIFFERENT METHODS OF PROTECTING MICROGRIDS

A valid security plot for the microgrid is to be structured so that in case of any deficiency, a base section of the framework is separated without affecting the rest of the framework. This should be possible with a mix of essential and back-up defensive equipment. The defensive equipment required to identify any defect inside its field.

In order to work, reinforcement at that point gets into equipment activity. The simple framework is spiral, and a subsequent security plot is planned with current transfers and wires that are very basic. As discussed before, with the introduction of distributed generations in the current framework, the structure of force becomes progressively more complex. Additionally, the flawed stream is unique because it varies with the method of activity, the number of the director-general and the

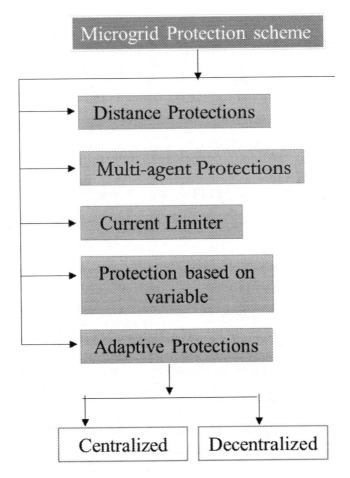

FIGURE 11.8 Organization followed for microgrid protection.

director-general. Subsequently, planning a productive insurance plan to adapt to the previously mentioned issues is a difficult task. To organize different insurance plans (Figure 11.8), an equivalent has been discussed.

11.5.1 Distance Protection

Remote protection that offers high selectivity applies significantly to transmission lines. The impedance equation is used to determine the train hand-off point

$$Z_m = \frac{Vot_m}{i_m} \tag{11.4}$$

where Vot_m is the calculated voltage and the calculated current at the transfer opinion. Under normal circumstances, intentional impedance incorporates heap impedance,

Security and Protection Techniques for Microgrids

and subsequently the value is higher. In any case, in the event that there is a deficiency in the line, the intentional impedance in such cases at that point would be same as the route impedance, which is extremely small. Thus, to determine intentional impedance and value, the occurrence of defects and what happened, deficiencies within the specific insurance sector are recognized. Generally, three areas of protection are advertised for a district. Zone 1 insurance is transient protection covering 80% of the area. Zone 2 insurance covers the area revealed through region 1. Region 3 provides insurance for a nearby link.

Both zones 2 and 3 work in clear time, and the time of activity of zone 3 is more noteworthy than that of zone 2. This priority is picked so that Z_{S1}, Z_{S2}, Z_{S3} The set worth is determined as

$$Z_{S1} = k_{r1} \times Z_{line} \tag{11.5}$$

$$Z_{21} = k_{r2} \times (Z_{line} + Z_{S1}) \tag{11.6}$$

$$Z_{S3} = k_{r3} \times (Z_{line} + Z_{S2}) \tag{11.7}$$

where k_{r1}, k_{r2}, k_{r3} are the risk coefficients of each zone, for the most part, set as an incentive between 0.8 and 0.85.

For distance reduction, a few steps applied with the microgrid (Figure 11.9) additionally add current to the distributed generation associated with bus 3. Director-general impedance will similarly generate line impedance, with the impedance seen from line 23 having high contrast with the impedance on bus 2. The impedance estimated by transfer $\Re_{23}$ is

$$Z_m = \frac{Vot_m}{i_m} = \frac{i_1 Z_{line23} + i_3 Z_\kappa}{i_1} \tag{11.8}$$

The intentional impedance set is transmitted by an incentive according to Eq. (8), and insurance selectivity is needed. Therefore, the estimated impedance in the upstream/downstream system is disturbed by the idea of distributed generation with a wide

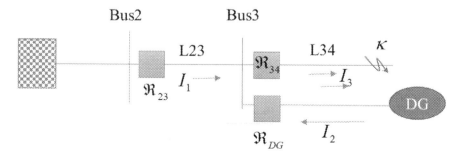

FIGURE 11.9 Distance protection.

range. To decrease the effect of the in feed current by distributed generation in the insurance, the set estimation of the transfers is recalculated as

$$Z_{S1} = \kappa_{r1} \times Z_{line} \tag{11.9}$$

$$Z_{21} = \kappa_{r2} \times (Z_{line} + \kappa_\beta Z_{S1}) \tag{11.10}$$

$$Z_{S3} = \kappa_{r3} \times (Z_{line} + \kappa_\beta Z_{S2}) \tag{11.11}$$

The new transfer settings are associated with great selectivity and efficacy and make assurance plans to work successfully for islanded mode.

Elements in the hand-off setting are used to ensure segregation with the aid of a focal controller due to occasional changes occurring in the system dependent on the position of the distributed generation and circuit breaker. The setting for each conceivable design (contingent on the state of the distributed generation and CB) is disconnected and put into an activity table.

11.5.2 Multi-Agent

This type of agent is a gathering of smart equipment and programming operators distributed in the system and works in combination to accomplish an overall objective, the effective security of the grid system. The engineering for a multi-specialist plot intended for versatile over-current assurance includes three layers: hardware, substation and framework. These layers are defined in Figure 11.9. A high range of current transfers is located in different pieces of the microgrid. The gear level is the most reduced level in a multi-expert organization and includes speculation operators, defender specialists, multipurpose specialists and entertaining specialists.

The measurement consultant screens the surrounding parameters (1 and 5) and takes them to the defender specialist. The defender operator checks to see if there is a problem with the nearby factors. At the time the event recognizes the subject, it sends data to the entertainment operator, and the entertainment operator trips the appropriate circuit breakers to disassemble the broken part.

The data of the multipurpose operator hardware works at the substation level. The substation level speaks through the gear level and moves to the data framework level. The substation level includes district specialists and executive operators. Region operators transmit the acquired data from portable experts to structure-level and board experts. The administration specialist determines the settings of the transitions to join the structure changes in the grid system and then hands it over to the defender operator. The most noticeable layer shapes the framework layer. The framework layer enters the regulator to organize the microgrid, speaks to the substation level and screens the entire system. It refreshes itself with system data that includes nearby factors, for example, the status of CBs, the status of distributed generations, electricity and power. In terms of these factors, it drives the mass flow and determines it theoretically.

Different types of parameters are sets for transfers and discusses it with substation level. The framework level includes a framework specialist and evaluation operator. The data obtained from the district expert is sorted by the framework operator,

Security and Protection Techniques for Microgrids

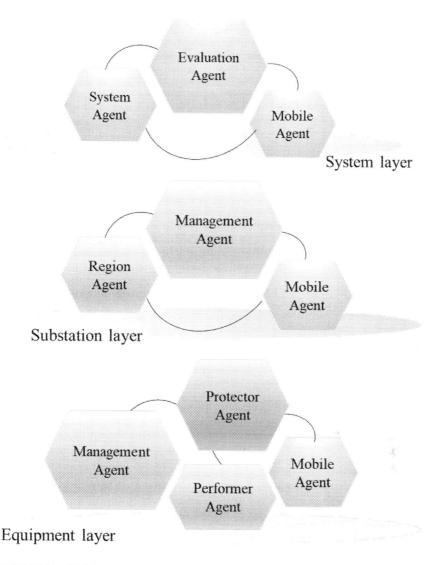

FIGURE 11.10 Multi-agent system.

and only the basic information is sent to the assessment operator. The evaluation expert reviews the assurance structure. In this event, it identifies any required insurance, instantly sending an adjusted setting to the administrative expert to accomplish smooth, multifaceted protection with the help of a multi-operator plot.

11.5.3 Current Regulator

Using the structure of a microgrid, fault current limitation is set up for the PCC to minimize the shortfall current. Fault current limiters (FCLs) are preserved at a minimum in normal conditions, and in most extreme situations, they are set below fault conditions [22]. Insufficient current limiting filters will be extensively sorted into three types: strong condition FCL, excellent indicator FCL and electromagnetic FCL. In the first, the electricity-minimizing behavior depends on the nonlinear response of the electric conducting materials to the temperature, electricity and area of interest [24]. Under normal conditions, the current stays below the edge; as a result, the whole current is given to defensive equipment. The currently restricted behavior in the powerful-state FCL is chosen depending on the on/off condition of the conductor equipment.

In the last type, current-minimizing behavior is chosen depending on the variety of attractive fields. The top deficit in the top half of the half-species is the current regulator [25], which is a mixture of superconducting FCL, strong-state FCL and electromagnetic FCL. For limiting current, this includes one of the proposed approaches in the microgrid system.

11.5.4 Variables Used in Protection

1. Microgrid security should be dependent on numerous factors, for example, current, voltage, edges, voicing waves, wavelet packet transformation (WPT) and THD tests that should be deliberated in detail, as follows.
2. Node powers. For the state factor, node voltages are used. The build-up between the intentional voltage and the state-rated voltage is extraordinary, but it indicates the proximity of the error; if not, the line is considered strong. State factors for the under-region are assessed. So the amount of correspondence required in this protection plot is lower.
3. WPT. Wavelet packet change is used to isolated, circulate and also limit all repetitive substances in the handled mark. It gives the signal in time and repeats limited signals which measure the repetition substance of the handled signal. When power electronic regulators are used, the three-phase sum is transformed to the edge of the DQ, and later the WPT is used for the signal. Measures of information of minimum detail length (MDL) are used to select the work of the ideal wavelet base and show that the Douchchies db4 waveform acts as the ideal.
4. Traveling wave. Traveling waves are separated by a mathematical morphological filter (MMF) that forms a clear element of magnification and disruptive signals. The two waves are separated and their time and intensity are kept separate. In the event that there is a fault in the first half, that point result of the poles of the two waves has a positive value, and in the case that it is contained in the second half, it will be negative.

11.5.5 Adaptive Protection

Adaptive protection is "an online movement that changes the preferred defensive response to adjustment in the position of the framework." A general plan for a versatile security framework has been introduced. As indicated by this arrangement, the

Security and Protection Techniques for Microgrids 249

assurance framework constantly screens the framework and implements new insurance coordination if any adjustment in topology or activity should occur. Prerequisites are required for this system,

1. Using directional and DOC transfers.
2. DOC must have cluster settings on the current transfer. These aggregated settings are changed locally or remotely.
3. Using a fast and secure informational foundation in a system of open conferences, such as IEC61850.

Adaptive protection can be made real and larger by a centralized and decentralized structure or two structures of multi-operators. In the two structures, the defensive sets of the microgrid should be refreshed in the event of any adjustment in the system. This is on the web or disconnected.

In the disconnected arrangement, each setting is simply cut into the memory of the experts or supervisory remote control unit (SRCU) for a dissimilar arrangement of the microgrid, and it is given a mode. SRCUs or specialist examinations use neighborhood information or take examples through an open framework. This activity is done online after each adjustment in the system. Data related to organized change help SRCUs or operators use related settings and use them as outing settings as long as the status does not exist.

The creators of [26, 27] tested this technique for two reasons. First, they claim that this strategy cannot cover all situations due to the various unique situations that exist in microgrids. They argue that this multifaceted strategy cannot evaluate synchronous problems because it requires a large amount of information augmentation due to the high number of possible situations to be measured.

11.5.6 CENTRALIZED DEFENSE SCHEME

This configuration uses an SRCU that is suitable for communication with transmission substation equipment through a fast and secure open infrastructure and transmits the necessary control and protective requests to equipment that monitors the situation and system conditions in general. The system should be provided to the SRCU as input, including the number and name, the load measurement existing in each transport and the properties of the force switches.

According to the arrangement introduced in [28], this problem is due to the constant change of microgrids. Despite this investigation, use online considerations for insurance adjustment [29]. In these investigations, if a distributed generation cooperates with or separates from the system, a regulator signal is directed to the SRCU to make new assumptions on the web and sets the electricity of each transfer activity according to the present condition [30].

11.5.7 DECENTRALIZED DEFENSE SCHEME

In opposition to focal versatile insurance, decentralized versatile assurance or multi-operators have advanced speed, consistent quality and expansibility, contrasting with brought-together security. Decentralized versatile assurance consists of many disseminated specialists in various equipment in the system that work in the whole organization, and can communicate through an informative framework and are consistent with neighborhood and wide-zone information [31]. For the most part, a specialist framework incorporates:

1. Autonomy: a specialist examines work conditionally and without the impedance of humans or different operators, per nearby circumstances.
2. Reactivity: specialist filters find modifications in input information and respond according to them.
3. Pre-liveliness: the specialist response includes information in a manner that fulfills the reasons for their structure.

11.6 DISCUSSION

The important parts exist in microgrid and for which, protection required are those-distributed generation, line, load and PCC. The modifier presence in the PCC should be planned with different security transfers and PCC current over, under-voltage, over-voltage, under-and over-repetitive transfers. The most suitable protection for microgrid is primary security with extra current as back up. However, by evaluating the various security mechanism aids designed for microgrids, it is recommended that multidisciplinary insurance is the best insurance. As a result, the versatile balance and scattering lines of different insurance frames are the best insurance scheme compared to the current protection for microgrid loads. The cost to implement versatile protection plots with advanced transfers and sensors on the raw edge is extremely high, but this cost is matched with required energy costs and lastly if this opportunity considers for long duration, then it provides better result.

11.7 FUTURE OF MICROGRIDS

Microgrids are important to the world and the future. We should examine some points demonstrated in Figure 11.11.

- They expand the effectiveness of bigger power grids.
- They offer force flexibility against catastrophic events, for example, seismic tremors, waves and storms.
- They help focus catastrophe aid during provincial and national emergencies.
- They are secure against digital and physical assaults.
- They are beneficial for the earth since they use renewable assets.
- They create job and training openings.

11.8 CONCLUSION

Today's world demand high-end development with various application of distributed technology. Higher requirement for distribution technology needs higher protection. This chapter dissects the techniques proposed in various investigations to take care of

Security and Protection Techniques for Microgrids

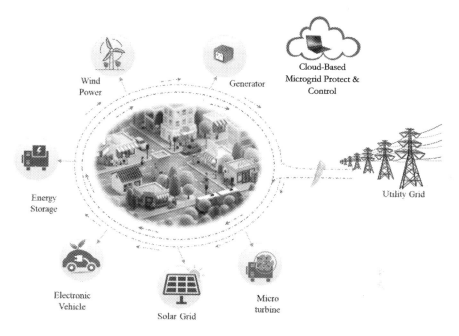

FIGURE 11.11 An illustration of the future of microgrids.

the issue of insurance in a microgrid. In this chapter, some standard assurance plans are investigated, which are applicable in a dissemination framework with smaller-scale matrices. Because of the preparation of a strong framework and the innovations of wide region estimation and PC security, specialists have begun to use new strategies to structure insurance control arrangements of networks. It was found that the existence of distributed generation units and microgrid capability of operation in two normal and islanding modes may change fault current level while causing false tripping, blindness of protection, prohibition of automatic reclosing and unsynchronized reclosing. In such a manner, key variables for structuring a suitable assurance framework for microgrids are talked about thoroughly. These comprise microgrid type and topology, distributed generation type, message interface type and interval time, a technique for shortcoming identification and investigation, transfer type, issue type, strategy for establishing and utilization of transformers in the microgrid.

Thinking about these elements, the different strategies projected so far for microgrid security were discussed. The survey underlines that utilizing a disconnected choice strategy in microgrid security can sabotage a vigorous assurance plan. This is because of the way that, with disconnected choices, all imaginable microgrid activity in all topologies may not be remembered for the insurance plan. On the other hand, utilizing a multi-specialist strategy is seen as the main technique equipped for reacting to microgrid auxiliary changes with high adaptability and consistent quality. In this technique, one can utilize suitable correspondence connections to investigate all system changes on the web and choose the best defensive procedure as indicated by the microgrid topology. Although, the vulnerabilities in correspondence joins and defensive hardware are seen as a challenge for this technique. A multipurpose insurance framework, applying new advancements, for example, PMU and GPS, a strong assurance framework and so on will be the future course of properly arranged security.

REFERENCES

1. Basak, P., et al., "A Literature Review on Integration of Distribute Energy Resources in the Perspective of Control, Protection and Stability of Microgrid," *Renewable and Sustainable Energy Reviews*, 16 (2012): 5545–5556.
2. Chen, J., et al., "The Overview of Protection Schemes for Distribution Systems Containing Micro-Grid," Proceedings of Power and Energy Engineering Conference, 2011, pp. 1–4.
3. Conti, S., "Protection Issues and State of the Art for Microgrids with Inverter Interfaced Distributed Generators," Proceedingsof International Conference on Clean Electrical Power (ICCEP), 2011, pp. 643–647.
4. Shanglin, Zhao, Wu Zaijun, Hu Minqiang, et al., "Thoughts on the Protection of Distributed Generation and Piconet," *Power System Automation*, 34.1 (2010): 73–77.
5. Che, L., et al., "Adaptive Protection System for Microgrids: Protection Practices of a Functional Microgrid System," *IEEE Electromagnetic*, 2 (2014): 66–80.
6. Shahriari, S. A. A., M. Abapour, A. Yazdian, et al., "Minimizing the Impact of Distributed Generation on Distribution Protection System by Solid State Fault Current Limiter," Transmission and Distribution Conference and Exposition, 2010 IEEE PES, 2010,pp. 1–7.
7. Nikkhajoei, Hassan, and Robert H. Lasseter, "Microgrid Protection," 2007 IEEE Power Engineering Society General Meeting, IEEE, 2007.
8. Hooshyar, Ali, and Reza Iravani, "Microgrid Protection," *Proceedings of the IEEE*, 105.7 (2017): 1332–1353.
9. Brearley, Belwin J., and R. Raja Prabu, "A Review on Issues and Approaches for Microgrid Protection," *Renewable and Sustainable Energy Reviews*, 67 (2017): 988–997.
10. Shiles, J., et al., "Microgrid Protection: An Overview of Protection Strategies in North American Microgrid Projects," 2017 IEEE Power & Energy Society General Meeting, IEEE, 2017.
11. Gururani, Ashika, Soumya R. Mohanty, et al., "Microgrid Protection Using Hilbert–Huang Transform Based-Differential Scheme," *IET Generation, Transmission & Distribution*, 10.15 (2016): 3707–3716.
12. Khandare, Pooja, S. A. Deokar, and A. M. Dixit, "Advanced Technique in Micro Grid Protection for Various Fault by Using Numerical Relay," 2017 2nd International Conference for Convergence in Technology (I2CT), IEEE, 2017.
13. Nuñez-Mata, Oscar, et al., "Development of a Microgrid Protection Laboratory Experiment for the Study of Overcurrent and Undervoltage Functions," 2017 CHILEAN Conference on Electrical, Electronics Engineering, Information and Communication Technologies (CHILECON), IEEE, 2017.
14. Habib, Hany F., Christopher R. Lashway, et al., "A Review of Communication Failure Impacts on Adaptive Microgrid Protection Schemes and the Use of Energy Storage as a Contingency," *IEEE Transactions on Industry Applications*, 54.2 (2017): 1194–1207.
15. Kar, Susmita, S. R. Samantaray, and M. Dadash Zadeh, "Data-Mining Model Based Intelligent Differential Microgrid Protection Scheme," *IEEE Systems Journal*, 11.2 (2015): 1161–1169.
16. Memon, Aushiq Ali, and Kimmo Kauhaniemi, "A Critical Review of AC Microgrid Protection Issues and Available Solutions," *Electric Power Systems Research*, 129 (2015): 23–31.

17. Cuzner, Robert M., and GiriVenkataramanan, "The Status of DC Micro-Grid Protection," 2008 IEEE Industry Applications Society Annual Meeting, IEEE, 2008.
18. Laaksonen, HannuJaakko, "Protection Principles for Future Microgrids," *IEEE Transactions on Power Electronics*, 25.12 (2010): 2910–2918.
19. Salomonsson, Daniel, Lennart Soder, and Ambra Sannino, "Protection of Low-Voltage DC Microgrids," *IEEE Transactions on Power Delivery*, 24.3 (2009): 1045–1053.
20. Ustun, TahaSelim, CagilOzansoy, and AladinZayegh, "Modeling of a Centralized Microgrid Protection System and Distributed Energy Resources According to IEC 61850-7-420," *IEEE Transactions on Power Systems*, 27.3 (2012): 1560–1567.
22. Kar, Susmita, and Subhransu Rajan Samantaray, "Time-Frequency Transform-Based Differential Scheme for Microgrid Protection," *IET Generation, Transmission & Distribution*, 8.2 (2014): 310–320.
23. Ustun, Taha Selim, Cagil Ozansoy, et al., "A Microgrid Protection System with Central Protection Unit and Extensive Communication," 2011 10th International Conference on Environment and Electrical Engineering, IEEE, 2011.
24. Oudalov, Alexandre, and Antonio Fidigatti, "Adaptive Network Protection in Microgrids," *International Journal of Distributed Energy Resources*, 5.3 (2009): 201–226.
25. Olivares, D. E., et al., "Trends in Microgrid Control," *IEEE Trans. Smart Grid*, 5.4 (July 2014): 1905–1919.
26. Al Lee, G., and W. Tschudi, "Edison Redux: 380 Vdc Brings Reliability and Efficiency to Sustainable Data Centers," *IEEE Power & Energy Magazine*, 10.6 (November 2012): 50–59.
27. Interconnection for Wind Energy, United States of America Federal Energy Regulatory Commission, Docket No. RM05-4-001, Order No. 661-A, December 2005. Available: www.ferc.gov/EventCalendar/ Files/20051212171744-RM05-4-001.pdf
28. Zeineldin, H. H., Y. A. R. I. Mohamed, V. Khadkikar, and V. R. Pandi, "A Protection Coordination Index for Evaluating Distributed Generation Impacts on Protection for Meshed Distribution Systems," *IEEE Trans. Smart Grid*, 4.3 (September 2013): 1523–1532.
29. Line Distance Protection REL670 Application Manual, Document 1MRK 506 315-UENABB, Vasteras, Sweden, December 2012. Available: www.05.abb.com/global/scot/ scot387.nsf/veritydisplay/1d6bacf573307830 c1257b0c0046292a/$file/1 MRK506315-UEN_C_en_Application_manual__REL670_1.2.pdf
30. Ziegler, G., *Numerical Distance Protection: Principles and Applications*, 4th ed. Erlangen, Germany: Publicis, 2011, ch. 6.
31. Ustun, Taha Selim, and Reduan H. Khan, "Multiterminal Hybrid Protection of Microgrids Over Wireless Communications Network," *IEEE Transactions on Smart Grid*, 6.5 (2015): 2493–2500.

12 Multilevel Voltage-Based Coordinating Controller Modeling, Development, and Performance Analysis

Ranjit Singh Sarban Singh, Tiara Natasya Abdul Halim, T. Joseph Sahaya Anand, and Maysam Abbod

CONTENTS

12.1	Introduction and Review of Chapter	255
12.2	System Integration: Multilevel Voltage-Based Coordinating Controller	256
	12.2.1 Electronic Stateflow Chart Circuit	257
	12.2.2 Electronic Logic AND Gate Circuit	259
	12.2.3 Electronic Conditional Switching Circuit	259
	12.2.4 Current Conversion Circuit	259
12.3	Results and Discussion	262
	12.3.1 Solar and Wind State flow Chart Outputs	262
	12.3.2 Condition Two: 12 V $\geq V_{solar} \leq$ 15 V and 10 V $\geq V_{wind} <$ 12 V	263
	12.3.3 Condition Three: 10 V $\leq V_{solar} <$ 12 V and 12 V $\leq V_{wind} \leq$ 15 V	266
	12.3.4 Condition Four: 0 V $< V_{solar} <$ 10 V and 0 V $< V_{wind} <$ 10 V	271
12.4	Conclusion	275
12.5	Acknowledgments	276
References		276

12.1 INTRODUCTION AND REVIEW OF CHAPTER

The irregularity of energy production and generation by renewable energy sources connected as a microgrid system makes it necessary to integrate an energy storage system (ESS) to stabilize the overall performance of the DC-based MG system. The evaluation of DC-based MG systems and their complexity made scholars introduce

DOI: 10.1201/9781003121626-12

energy management systems (EMSs) (Zia et al., 2018) (Anvari-Moghaddam et al., 2017; Karavas et al., 2015; Kofinas et al., 2018; Liu et al., 2018; Yoon et al., 2018). EMS development and technology for MG-based systems can be categorized into centralized EMS (CEMS) (Anvari-Moghaddam et al., 2017) and decentralized EMD (DEMS) (Liu et al., 2018). CEMS is an AC-based system connected to MG grid transmission, while DEMS is an MG DC-based system, known as a standalone system.

Employment of EMS in a DC-based MG system for distributed generation sources such as solar PV systems, wind turbine systems, hydropower systems and so on provides an ideal and reliable solution to continuously supply electricity power to rural communities where the national transmission grid is not reachable (Chen et al., 2018).

The proposed and presented systems in Anvari-Moghaddam et al. (2017), Salgueiro et al. (n.d.) and Moghaddam (2018) deploy a centralized agent/controller where it transfers or shares electricity information with other connected agents. The other connected agents are RESs such as solar and wind, the national transmission grid, load and ESS. The centralized agent/controller acts as the master controller and is responsible for managing the data as well as sending instructions to the other connected agents based on real-time information received by the master agent/controller. In Karavas et al. (2015), a decentralized multi-agent system was introduced. Each agent in the proposed decentralized multi-agent system operates when information is received through connected sensors. The disadvantage of the decentralized multi-agent system is that each agent has its own algorithm and needs to be synchronized with complicated coding for a smooth functionality and operation. In another related study, Ranjit et. al., Singh et al. (2015) and Singh and Abbod (2019) proposed and developed a voltage divider-switching technique to sense and measure the output voltages of RES in a decentralized hybrid renewable energy system. The developed voltage divider-switching configuration is proposed to perform self-intervention between the solar PV and wind energy systems of hybrid renewable energy systems. The voltage divider-switching technique successfully manages to simultaneously coordinate and control the supply to load and battery storage charging/discharging.

Through the presented advantages and disadvantages of EMS in AC- and DC-based MG systems, this chapter proposes to develop and model an EMS DC-based system using a multilevel voltage-based coordinating controller. The designed and modeled multilevel voltage-based coordinating controller simulation and analysis would be an advantage to develop a hybrid RES system which can be used for rural community implementation. This chapter also presents the multilevel voltage-based coordinating controller's design and overall system operation, which will be used to validate the system's ability to sense and measure, switch and effectively coordinate and manage the produced output voltage from a solar PV system and wind energy system for load supply or/and energy storage charging/discharging.

12.2 SYSTEM INTEGRATION: MULTILEVEL VOLTAGE-BASED COORDINATING CONTROLLER

The multilevel voltage-based coordinating controller is responsible for sensing and measuring, switching, effectively coordinating and managing the distributed generated energies from solar PV and wind energy system renewable energy sources. The importance of switching, effectively coordinating and managing is to simultaneously perform different roles with the same objective to supply the distributed generated energy to the connected load. Figure 12.1 shows the block diagram of the multilevel voltage-based coordinating controller integrated system. This system integrates two RESs, solar energy from a solar PV system and wind energy from a wind energy system. Analog output voltages from these RESs are measured and sent to the electronic stateflow condition circuit.

The measured output voltages are categorized into three parameters: $0V \leq V_R < 10V$, $10V \leq V_R < 12V$ and $12V \leq V_R \leq 15V$ where R is the RESs used to simulate the developed model. The proposed multilevel voltage-based coordinating controller model is built based on nine combinational voltage conditions, as shown in Table 10.1. These nine combinational voltage conditions are placed in the electronic condition switching circuit sub-system.

The electronic stateflow condition circuit is used to sense and measure the output voltages from RESs. The electronic stateflow condition circuit is composed with two stateflow sub-sub-systems, which sense and measure the output voltages of the RESs. Each stateflow produces a digital 0/1 value which represents the sensed and measured input voltage from the RES. The 0/1 digital value produced by the stateflow sub-sub systems, which is the combination of two voltages from Table 12.1, is then sent into the electronic logic AND gate circuit sub-system. The 0/1 single logic output from the electronic logic AND gate circuit is used to switch the respective circuitry in the electronic condition switching circuit, which is responsible for controlling the role of the RESs. In other words, the electronic condition switching circuit manages the produced energy source from the RESs to supply the connected load or/and charge the battery energy storage system.

12.2.1 Electronic Stateflow Chart Circuit

The electronic stateflow chart circuit simultaneously senses and measures the output voltages from RESs, which helps to effectively coordinate the switching in the electronic switching condition circuit. The electronic stateflow chart circuit consists of the solar and wind stateflow chart, which is based on multilevel voltages. The solar and wind stateflow chart design is shown in Figure 12.2.

The solar and wind stateflow chart acts as the coordinator mechanism which decides which RES will be connected to the load and battery or only the battery for charging. This helps the multilevel voltage-based coordinating controller make quick decisions on effective directive coordination. As mentioned in Section 12.2, the sensed and measured voltages through the solar and wind stateflow chart will produce logic 0/1 output. Hence, the solar and wind stateflow chart-produced logic 0/1 is then sent to the electronic logic AND gate circuit. At the electronic logic AND gate circuit, these two logics are combined to output one single logic, which is used to control the switching of the electronic switching condition circuit. Referring to Figure 12.2, solar and wind stateflow charts are divided into multilevel voltage paths or flow parameters (solar0, solar10, solar12, wind0, wind10 and wind12). Thus, when there is an input voltage from solar or wind, the solar and wind stateflow chart determines which path or flow the input voltage should take.

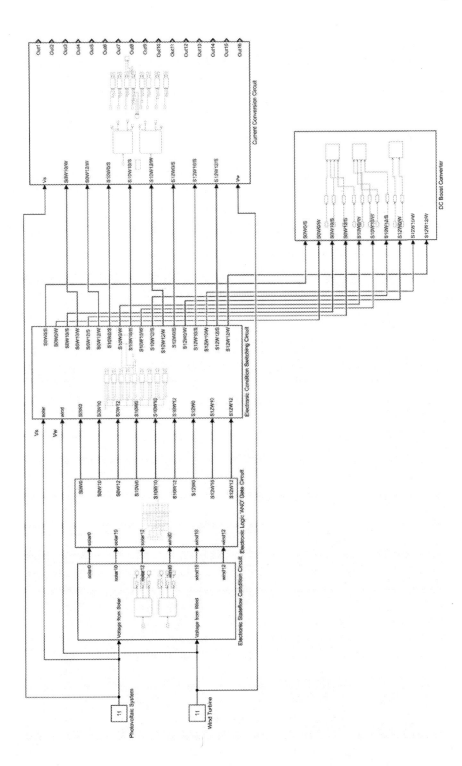

FIGURE 12.1 Multilevel voltage-based coordinating controller.

TABLE 12.1
Nine Combinational Voltage Conditions

No	Combinational Voltage Conditions
1	$0\ V \leq V_{solar} < 10\ V$ and $0\ V \leq V_{wind} < 10\ V$
2	$0\ V \leq V_{solar} < 10\ V$ and $10\ V \geq V_{wind} < 12\ V$
3	$0\ V \leq V_{solar} < 10\ V$ and $12\ V \geq V_{wind} \leq 15\ V$
4	$10\ V \geq V_{solar} < 12\ V$ and $0\ V \leq V_{wind} < 10\ V$
5	$10\ V \geq V_{solar} < 12\ V$ and $10\ V \geq V_{wind} < 12\ V$
6	$10\ V \geq V_{solar} < 12\ V$ and $12\ V \geq V_{wind} \leq 15\ V$
7	$12\ V \geq V_{solar} \leq 15\ V$ and $0\ V \leq V_{wind} < 10\ V$
8	$12\ V \geq V_{solar} \leq 15\ V$ and $10\ V \geq V_{wind} < 12\ V$
9	$12\ V \geq V_{solar} \leq 15\ V$ and $12\ V \geq V_{wind} \leq 15\ V$

12.2.2 Electronic Logic AND Gate Circuit

Figure 12.3 shows the electronic logic AND gate circuit integrated inside the multilevel voltage-based coordinating controller model. The electronic logic AND gate circuit is a combination of AND gates which takes in two output logics from the solar and wind stateflow chart, as explained in the previous section. The single logic output from the electronic logic AND gate circuit will determine any two conditions, as presented in Table 12.1, for the sub-system electronic switching condition circuit to switch on and connect to the connected load or battery for charging.

The six parameters shown in the solar and wind stateflow chart in Figure 12.2 are mapped to all possible outputs to optimize the developed multilevel voltage-based coordinating controller model utilization. Hence, the nine combinational voltage conditions shown in Table 12.1are presented in Figure 12.4.

12.2.3 Electronic Conditional Switching Circuit

The electronic conditional switching circuit is built based on the controlled voltage source and input relay switching based on the logic condition from the electronic logic AND gate circuit. There are nine sub-sub-systems built in the electronic conditional switching circuit sub-system, which are based on the nine combinational voltage conditions shown in Table 12.1. One of the sub-sub-system components is shown in Figure 12.5. Voltage from solar and wind is connected to the solar and wind inputs when S0W0 receives a 1 logic from the electronic logic AND gate circuit. The relays at S0W0 are turned on which will switch on the SPDT relays connected to the controlled voltage source. When the SPDT relays are switched on, the S and W ports will be connected either to the connected load or battery energy storage system for charging.

12.2.4 Current Conversion Circuit

The modeled multilevel voltage-based coordinating controller carries DC source supply from the RESs before the DC source is converted into an AC source for the

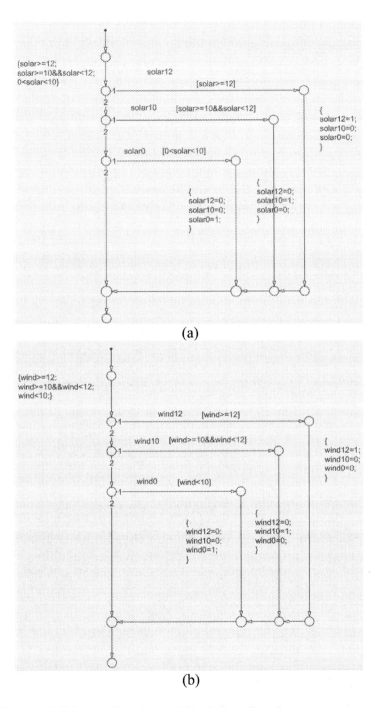

FIGURE 12.2 (a) Solar stateflow chart and (b) wind stateflow chart.

Multilevel Voltage-Based Coordinating Controller

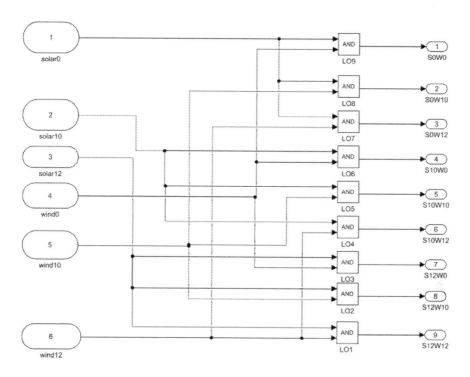

FIGURE 12.3 AND gate combinational logics in electronic logic AND gate circuit.

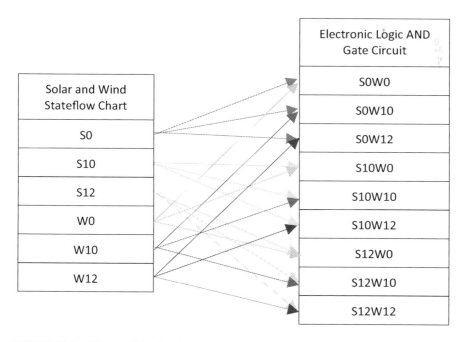

FIGURE 12.4 Nine combinational voltage conditions derived from solar and wind stateflow chart.

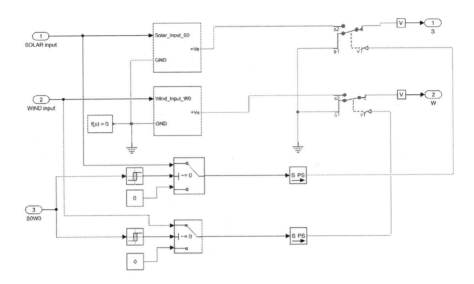

FIGURE 12.5 Controlled voltage source switching circuit.

connected load. To perform the conversion, the current conversion circuit is integrated to convert the current of the output voltage before supplying the voltage to the connected load. DC variable input voltage ranging from 10-V DC to 15-V DC is received in a boost converter to step up the low DC voltage to 15 V before converting to AC supply. The 10-V DC ≥ volt ≥ 14-V DC voltages are used to converter the low DC voltages into 15-V DC before connecting them to a DC-to-AC inverter for 240-V AC output.

12.3 RESULTS AND DISCUSSION

The proposed multilevel voltage-based coordinating controller has been built using MATLAB/Simulink software. The multilevel voltage-based coordinating controller is integrated with all the sub-systems that have been described, as well as with the boost converter, inverter and path to battery energy storage system. The boost is integrated at the different sub-systems: (1) current conversion circuit and (2) DC boost converter sub-systems. The boost converter integrated in the DC boost converter sub-system receives either one of the low 10-V DC ≥ volt ≥ 14-V DC RES output voltage to be stepped up to DC 15-V for battery energy storage system charging.

The results of the modeled multilevel voltage-based coordinating controller are presented in this section. A few conditions are selected from the conditions presented in Table 12.1.

12.3.1 Solar and Wind State flow Chart Outputs

Table 12.2 presents the operating condition results for the solar and wind stateflow chart shown in Figure 12.2.

Multilevel Voltage-Based Coordinating Controller

TABLE 12.2
Solar and Wind Stateflow Chart Logic Output Conditions

	Output Voltage - Solar	solar0	solar10	solar12
Solar Stateflow Chart	$0V \leq V_{solar} < 10V$	1		
	$10V \geq V_{solar} < 12V$		1	
	$12V \geq V_{solar} \leq 15V$			1

	Output Voltage - Wind	wind0	wind10	wind12
Wind Stateflow Chart	$0V \leq V_{wind} < 10V$	1		
	$10V \geq V_{wind} < 12V$		1	
	$12V \geq V_{wind} \leq 15V$			1

Condition One: $12\ V \geq V_{solar} \leq 15\ V$ and $12\ V \geq V_{wind} \leq 15\ V$

In condition one, the output voltages V_{solar} and V_{wind} from the RESs are 15-V DC, which is categorized as $12\ V \leq V_R \leq 15\ V$. For both conditions, solar12 and wind12 produce logic 1, as shown in Figure 12.6. As mentioned in Section 12.2.1, when condition one occurs, both the solar and wind stateflow charts will produce a logic 1, and this logic 1 is sent into the AND gate combinational logics in the electronic logic AND gate circuit shown in Figure 12.3 through solar12 (port 3) and wind12 (port 6). Figure 12.7 shows the S12W12 AND gate output logic result at S12W12 port 9 (Figure 12.3). The logic 1 shown in Figure 12.7 be also can be measured at port S12W12 shown in Figure 12.1, which is the input into the electronic switching condition circuit shown in Figure 12.5, replacing S0W0 to S12W12.

Figure 12.8 shows the solar and wind 15-V DC voltage output at the electronic switching condition circuit, which is being supplied by the RESs. The solar–15-V DC output voltage at port S12W12/S is connected to the current conversion circuit, and the wind–15-V DC output voltage at port S12W12/W is connected to the DC boost converter, as shown in Figure 12.1.

Figure 12.9 shows the S12W12/S–15-V DC output voltage shown in Figure 12.8 sent to the current conversion circuit and a 240-AC output voltage produced for the connected loads. On the other hand, the S12W12/W–15-V DC output voltage is sent to the DC boost converter to directly start charging the battery energy storage system.

12.3.2 Condition Two: $12\ V \geq V_{SOLAR} \leq 15\ V$ and $10\ V \geq V_{WIND} < 12\ V$

In condition two, the output voltages V_{solar} and V_{wind} from RESs are 15-V DC and 11.5-V DC, which is categorized as $12\ V \leq V_R \leq 15\ V$ and $10\ V \leq V_R < 12\ V$. For both

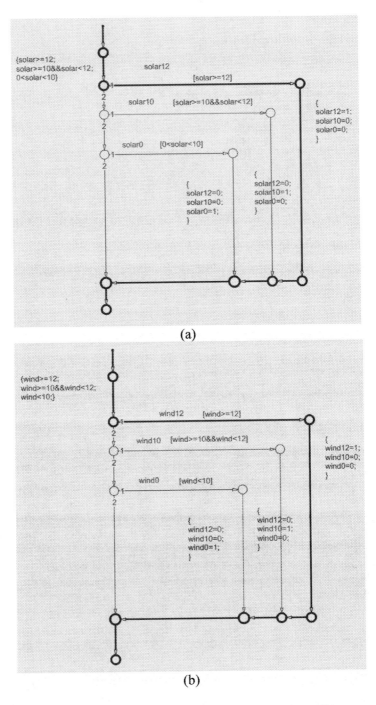

FIGURE 12.6 Solar12 and wind12 stateflow output voltage—logic conditions.

Multilevel Voltage-Based Coordinating Controller

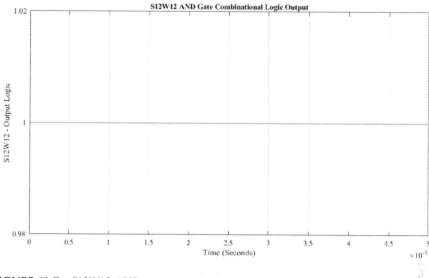

FIGURE 12.7 S12W12 AND gate output logic.

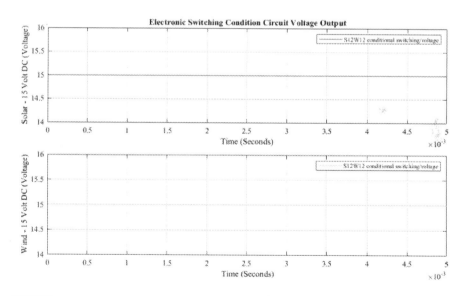

FIGURE 12.8 S12W12 controlled voltage source output—electronic switching condition circuit.

conditions, solar12 and wind10 produce logic 1, as shown in Figure 12.10. As mentioned in Section 12.2.1, when condition two occurs, both the solar and wind stateflow charts will produce a logic 1, and this logic 1 is sent to the AND gate combinational logics in the electronic logic AND gate circuit shown in Figure 12.3 through solar12 (port 3) and wind10 (port 5). Figure 12.11 shows the S12W10 AND gate output logic

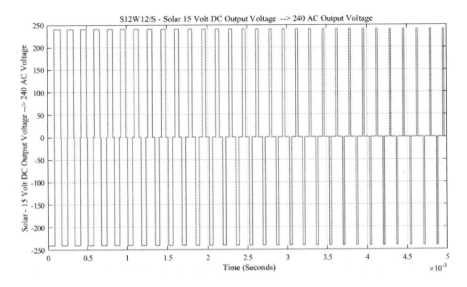

FIGURE 12.9 S12W12/S—15-V DC output voltage ◊ 240 AC output voltage.

result at S12W10 port 8 Figure 12.3. The logic 1 shown in Figure 12.11 can also be measured at port S12W10 shown in Figure 12.1, which is the input into the electronic switching condition circuit shown in Figure 12.5, replacing S0W0 to S12W10. Figure 12.12 shows the solar 15-V DC voltage output and wind 11.5-V DC voltage output at the electronic switching condition circuit which is being supplied by the RESs. The solar–15-V DC output voltage at port S12W10/S is connected to the current conversion circuit, and the wind–11.5-V DC output voltage at port S12W10/W is connected to the DC boost converter, as shown in Figure 12.1.

Figure 12.13 shows the S12W10/S–15-V DC output voltage shown in Figure 12.12 is sent into the current conversion circuit and a 240-AC output voltage is produced for the connected loads, whereas the S12W10/W–11.5-V DC output voltage is sent to the DC boost converter to produce a 13.21-V DC output voltage, as shown in Figure 12.14 for battery energy storage system charging.

12.3.3 Condition Three: $10 \text{ V} \leq V_{solar} < 12 \text{ V}$ and $12 \text{ V} \leq V_{wind} \leq 15 \text{ V}$

In condition three, the output voltages V_{solar} and V_{wind} from RESs are 11.5-V DC and 15-V DC, which is categorized as $10 \text{ V} \leq V_R < 12 \text{ V}$ and $12 \text{ V} \leq V_R \leq 15 \text{ V}$. For both conditions, solar10 and wind12 produce logic 1, as shown in Figure 12.15. As mentioned in Section 12.2.1, when condition two occurs, both the solar and wind stateflow charts will produce a logic 1, and this logic 1 is sent to the AND gate combinational logics in the electronic logic AND gate circuit shown in Figure 12.3 through solar10 (port 2) and wind12 (port 6). Figure 12.16 shows the S10W12 AND gate output logic result at S10W12 port 6 Figure 12.3. The logic 1 shown in Figure 12.16 can also be measured at port S10W12 shown in Figure 12.1, which

Multilevel Voltage-Based Coordinating Controller

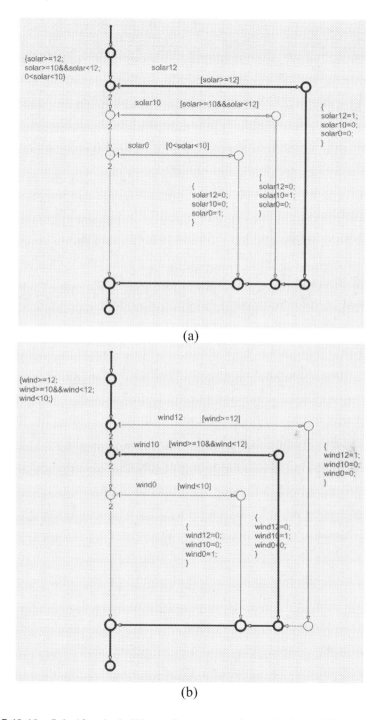

FIGURE 12.10 Solar12 and wind10 stateflow output voltage—logic conditions.

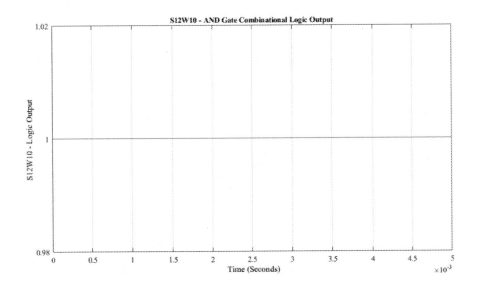

FIGURE 12.11 S12W10 AND gate output logic.

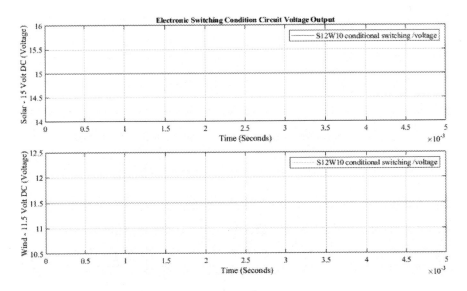

FIGURE 12.12 S12W10 controlled voltage source output—electronic switching condition circuit.

is the input into the electronic switching condition circuit shown in Figure 12.5, replacing S0W0 to S10W12. Figure 12.17 shows the solar 11.5-V DC voltage output and wind 15-V DC voltage output at the electronic switching condition circuit, which is being supplied by the RESs. The solar–11.5-V DC output voltage at port S10W12/S is connected to the DC boost converter, and the wind–15-VDC output

Multilevel Voltage-Based Coordinating Controller

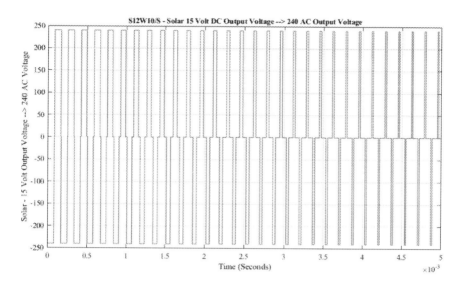

FIGURE 12.13 S12W10/S—15-V DC output voltage ◊ 240 AC output voltage.

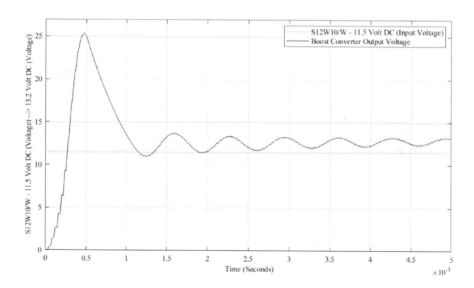

FIGURE 12.14 S12W10/W—11.5-V DC input voltage ◊ 13.21-V DC output voltage.

voltage at port S10W12/W is connected to the conversion current circuit, as shown in Figure 12.1.

Figure 12.18 shows the S10W12/W–15-V DC output voltage shown in Figure 12.17 is sent into the current conversion circuit and a 240-AC output voltage is produced for the connected loads, whereas the S10W12/S–11.5-V DC output voltage is sent into the DC boost converter to produce a 13.21-V DC output voltage, as shown in Figure 12.19, for battery energy storage system charging.

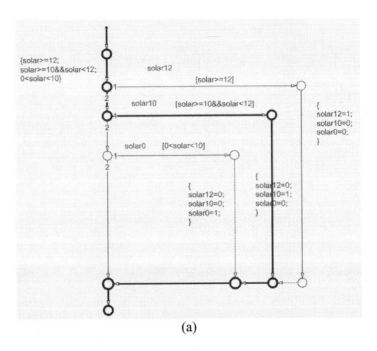

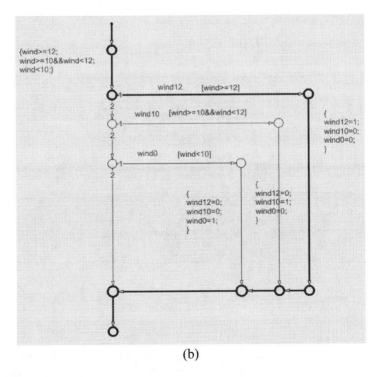

FIGURE 12.15 Solar10 and wind12 stateflow output voltage—logic conditions.

Multilevel Voltage-Based Coordinating Controller

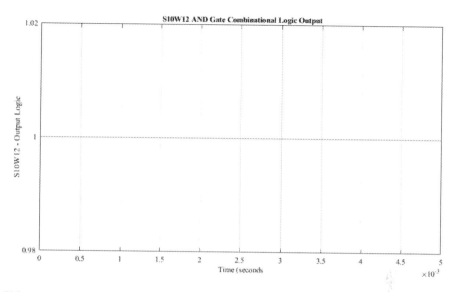

FIGURE 12.16 S10W12 AND gate output logic.

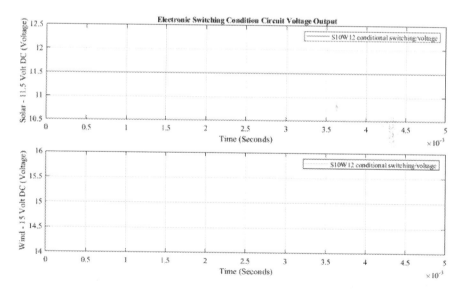

FIGURE 12.17 S10W12 controlled voltage source output—electronic switching condition circuit.

12.3.4 Condition Four: 0 V < V_{SOLAR} < 10 V and 0 V < V_{WIND} < 10 V

In condition three, the output voltages V_{solar} and V_{wind} from the RESs are 11.5-V DC and 15-V DC, which are categorized as 0 V < V_R < 10 V and 0 V < V_R < 10 V. For both conditions, the solar0 and wind0 produce logic 1, as shown in Figure 12.20. As

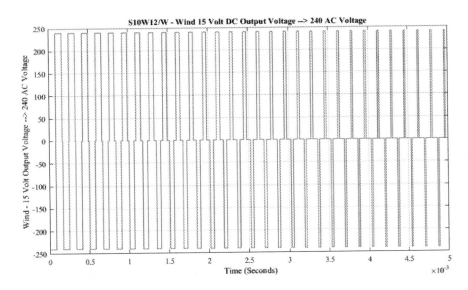

FIGURE 12.18 S12W10/S—15-V DC output voltage ◊ 240-AC output voltage.

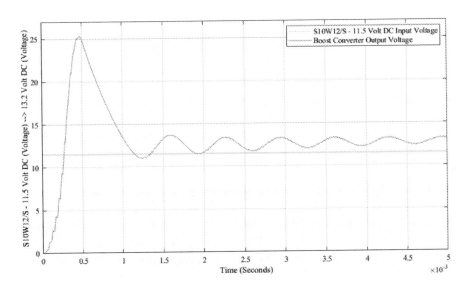

FIGURE 12.19 S10W12/S–11.5-V DC input voltage ◊ 13.21-V DC output voltage.

mentioned in Section 12.2.1, when condition two occurs, both the solar and wind stateflow charts will produce a logic 1, and this logic 1 is sent to the AND gate combinational logics in the electronic logic AND gate circuit shown in Figure 12.3 through solar0 (port 1) and wind0 (port 4). Figure 12.16 shows the S10W12 AND gate output logic result at S10W12 port 6 Figure 12.3. The logic 1 shown in Figure 12.21 can also

Multilevel Voltage-Based Coordinating Controller

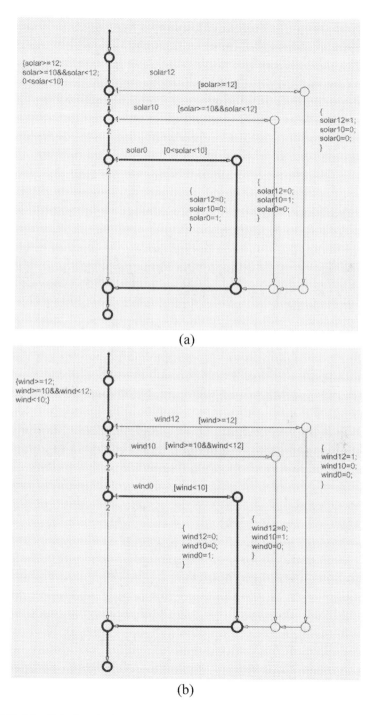

FIGURE 12.20 Solar0 and wind0 stateflow output voltage—logic conditions.

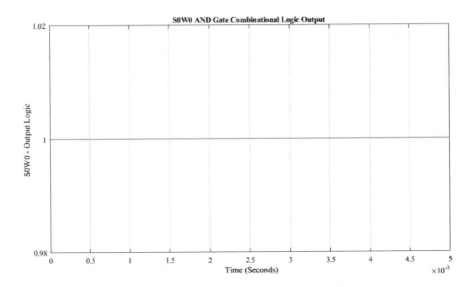

FIGURE 12.21 S0W0 AND gate output logic.

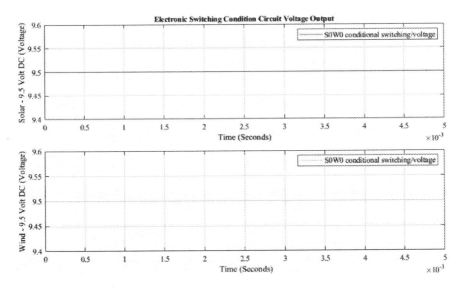

FIGURE 12.22 S0W0 controlled voltage source output—electronic switching condition circuit.

be measured at port S0W0 shown in Figure 12.1, which is the input into the S0W0 electronic switching condition circuit shown in Figure 12.5. Figure 12.22 shows the solar 9.5-V DC voltage output and wind 95-V DC voltage output at the electronic switching condition circuit, which is being supplied by the RESs. The solar and wind–9.5-V DC output voltage at ports S0W0/S and S0W0/W are connected to the DC boost converter, as shown in Figure 12.1.

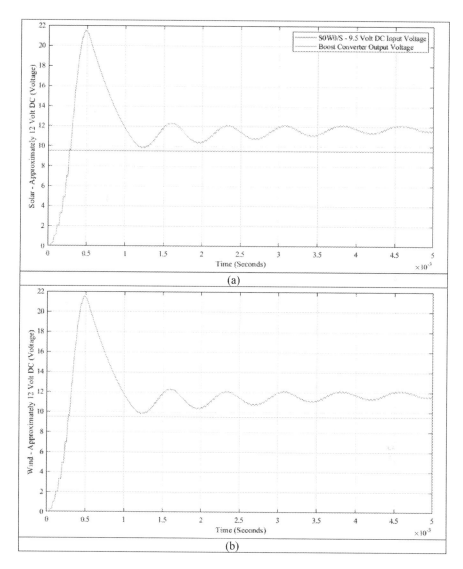

FIGURE 12.23 (a) S0W0/S–9.5-V DC input voltage ◊ 12-V DC output voltage and (b) S0W0/W–9.5-V DC input voltage ◊ 12-V DC output voltage.

Figure 12.23 shows the S0W0/S and S0W0/W input 9.5-V DC voltage from the RESs are sent into the DC boost converter to produce an approximately 12-V DC voltage output at the DC boost converter. This 12-V DC voltage output can be used for DC source appliances.

12.4 CONCLUSION

The multilevel voltage-based coordinating controller is proposed to sense and measure, switch, effectively coordinate and manage the generated energy from two renewable energy sources (solar and wind). The modeled and developed multilevel voltage-based coordinating controller has successfully demonstrated the objectives of sensing and measuring, switching, effectively coordinating and managing the generated energy from two renewable energy sources for either AC load supply via the conversion current circuit or battery energy storage system charging via the DC boost converter shown in Figure 12.1. All the captured results validate the modeled and developed multilevel voltage-based coordinating controller performance, and the produced results are also analyzed to indicate that the modeled and developed multilevel voltage-based coordinating controller is able to sense and measure, switch between the sources for energy supply or battery energy storage system charging, effectively coordinate the produced energy for the load and battery energy storage system charging and, last, manage the overall modeled and developed system performance without causing any failure.

12.5 ACKNOWLEDGMENTS

The authors gratefully acknowledge the support of Centre of Telecommunication Research & Innovation (CeTRI), Faculti Kejuruteraan Elektronik dan Kejuruteraan Komputer (FKEKK) and Sustainable and Responsive Manufacturing Group (SUSREM), Faculty of Manufacturing Engineering (FKP), Universiti Teknikal Malaysia Melaka, as well as the Department of Electronic and Computer Engineering, College of Engineering, Design and Physical Sciences, Brunel University London, UK.

REFERENCES

Anvari-Moghaddam, A., Rahimi-Kian, A., Mirian, M. S., & Guerrero, J. M. (2017). A Multi-Agent Based Energy Management Solution for Integrated Buildings and Microgrid System. *Applied Energy*, *203*, 41–56. https://doi.org/10.1016/j.apenergy.2017.06.007.

Chen, M., Ma, S., Wan, H., Wu, J., & Jiang, Y. (2018). Distributed Control Strategy for DC Microgrids of Photovoltaic Energy Storage Systems in Off-Grid Operation. *Energies*, *11*(10), 2637. https://doi.org/10.3390/en11102637.

Karavas, C. S., Kyriakarakos, G., Arvanitis, K. G., & Papadakis, G. (2015). A Multi-Agent Decentralized Energy Management System Based on Distributed Intelligence for the Design and Control of Autonomous Polygeneration Microgrids. *Energy Conversion and Management*, *103*, 166–179. https://doi.org/10.1016/j.enconman.2015.06.021.

Kofinas, P., Dounis, A. I., & Vouros, G. A. (2018). Fuzzy Q-Learning for Multi-Agent Decentralized Energy Management in Microgrids. *Applied Energy*, *219*(March), 53–67. https://doi.org/10.1016/j.apenergy.2018.03.017.

Liu, W., Li, N., Jiang, Z., Chen, Z., Wang, S., Han, J., Zhang, X., & Liu, C. (2018). Smart Micro-Grid System with Wind/PV/Battery. *Energy Procedia*, *152*, 1212–1217. https://doi.org/10.1016/j.egypro.2018.09.171.

Moghaddam, I. N. (2018). *Optimal Sizing and Operation of Energy Storage Systems to Mitigate Intermittency of Renewable Energy Resources*, ProQuest LLC, The University of North Carolina at Charlotte.

Salgueiro, Y., Rivera, M., & Gonzalo, N. (2019). *Multiagent-Based Decision Support Systems in Smart Microgrids*, Intelligent Decision Technologies, Smart Innovation, Systems and Technologies, Springer, Singapore.

Singh, R. S. S., & Abbod, M. (2019). Self-Assistive Controller Using Voltage Droop Method for DC Distributed Generators and Storages. In *Cases on Green Energy and Sustainable Development* (pp. 379–405). https://doi.org/10.4018/978-1-5225-8559-6.ch014.

Singh, R. S. S., Abbod, M., & Balachandran, W. (2015). Renewable Energy Resource Self-Intervention Control Technique Using Simulink/Stateflow Modeling. *Proceedings of the Universities Power Engineering Conference, 2015*(September). https://doi.org/10.1109/UPEC.2015.7339803.

Yoon, S., Kim, S., Park, G., Kim, Y., Cho, C., & Park, B. (2018). Multiple Power-Based Building Energy Management System for Efficient Management of Building Energy. *Sustainable Cities and Society, 42*(August), 462–470. https://doi.org/10.1016/j.scs.2018.08.008.

Zia, M. F., Elbouchikhi, E., & Benbouzid, M. (2018). Microgrids Energy Management Systems: A Critical Review on Methods, Solutions, and Prospects. *Applied Energy, 222*(April), 1033–1055. https://doi.org/10.1016/j.apenergy.2018.04.103.

13 Case Studies on Microgrid Design Using Renewable Energy Sources

Arjyadhara Pradhan, Babita Panda, and Rao Mannepalli

CONTENTS

13.1	Introduction	280
13.2	Necessity of Microgrids	280
	13.2.1 Brief about Microgrids	281
	13.2.2 Basic Components of Microgrids	281
	13.2.3 Challenges	281
13.3	Renewable Energy Sources: Overview	282
	13.3.1 Photovoltaic Standalone Systems	283
	13.3.2 Design of Simple DC Microgrid for AC Load	283
	13.3.3 MATLAB/Simulink	284
	13.3.4 Results and Discussion	286
13.4	Hybrid System Using HOMER Software	288
	13.4.1 Microgrid System Modeling: Case Study	290
	13.4.1.1 PV Modeling	291
	13.4.1.2 Wind Turbine Modeling	291
	13.4.1.3 Battery Modeling	291
	13.4.1.4 Diesel Generator Modeling	291
	13.4.1.5 Converter Modeling	291
	13.4.1.6 Utility Grid	291
	13.4.1.7 Load Profile	292
	13.4.2 Energy Management	292
	13.4.3 Simulation	293
13.5	Conclusion	296
References		297

DOI: 10.1201/9781003121626-13

13.1 INTRODUCTION

Microgrids provide democratic energy delivery. In reality and today's world, microgrids operate on diesel and other fuels which are not practical to support daily life activities and are just meant for short-term emergencies. Microgrid projects can better grow and show fruitful results with the deployment of renewable energy, as they are very cost effective and environmentally friendly [1]. Electrification and diversification of our energy portfolios to provide clean energy is the biggest step of each and every microgrid [2]. Let us consider a case study: a factory equipped with a diesel generator is a microgrid. Thus, when the grid loses power, the factory can operate normally until there is enough fuel for the distributed generation set to run, and the fuel used for the distributed generation set is costly and also creates pollution in the environment. Thus, microgrids using renewable sources of energy, like solar and wind, help generate clean energy and support the factory. With the development of storage battery technology, microgrids are today equipped with more batteries to withstand periods of high demand and low production [3]. Thus, they become more resilient.

Moreover, solar and wind energy are not constant, as sun can be blocked by clouds, and the speed of the wind is also not constant; it changes frequently. Hence the energy produced is not used completely, so providing buildings and homes with small grid-connected distributed energy resources is beneficial, as it reduces energy loss in the process of distributing and transmitting energy from any non-conventional sources like solar and wind [4].

13.2 NECESSITY OF MICROGRIDS

To provide a better habitat for human civilization, the energy industry must focus on total electrification of the transportation, residential and commercial sectors. Not only this, but an efficient and resilient power supply must be maintained by the power grid [5]. Among several renewable energy sources, solar and wind support major power production. But they are also inconsistent depending on weather conditions and different environmental factors. In this digital world where uninterruptible power is required, for example, for hospitals, these two clean forms of energy require assistance to prevent blackout. In hospitals, lives hang in the balance and many machines requires a constant source of supply for patients. Therefore, the power grid must ensure that the power from renewable sources, which is clean and eco-friendly, must be used where it is needed immediately and feeds excess electricity back into the system. This is possible by using microgrids.

There are many others reasons microgrids have come into the picture. Microgrids are offline standalone systems providing electricity to rural areas where basic infrastructure is not there and in some cases has been destroyed by natural disasters [6]. Apart from this, microgrids also help to produce power, so generating clients like residential and commercial buildings can have solar array installations on the rooftops. Thus, when surplus power is produced apart from their requirements, it can be fed back to the grid, providing a chance for income for the owners of the building. Moreover, traditional grids can no longer support the sustainable use of energy, as about 5% of power is wasted by transporting energy over distances. Microgrids allow generation from renewable sources without undergoing any redesign of the grid system.

13.2.1 Brief about Microgrids

A microgrid is a local energy grid with its own control capability to operate both with the grid and also autonomously. Traditional grids connect residential and commercial buildings to central power sources, allowing the use of appliances, electronics and so on. This shows that, as all the systems are interconnected, whenever the grid requires maintenance or any problem occurs, the total system is affected [7]. But in the case of microgrids, it has the advantage to cut and self-operate using its own energy generation during natural disasters or any kind of power outage. There are different ways of powering a microgrid, like distributed generators, batteries and even renewable sources.

Microgrids are connected to the grid at a point of common coupling that maintains voltage at the same level as the main grid [8]. A separate switch is used to separate the microgrid and main grid. Thus, the microgrid can act as an island. Microgrids cut costs and mostly use local resources, which conventional grids would not use.

13.2.2 Basic Components of Microgrids

There are four basic components of microgrids: local generation, consumption, energy storage and point of common coupling.

A microgrid consists of different sources that supply electricity to the end user. These sources are categorized into two major areas, thermal energy sources (natural gas, micro combined heat and power) and renewable sources (solar, wind, bio). In terms of consumption, any elements which consume electricity, like lighting and heating systems of buildings, both residential and commercial, can be connected with the microgrid [9]. In special cases with controllable loads, consumption patterns can be modified in terms of demand. In microgrids, energy storage performs a lot of functions like maintaining proper power quality, regulation of voltage and frequency, smoothing output of renewable energy sources, provision of backup power and cost optimization. This energy storage can be of various types, like chemical, electrical, gravitational, flywheel and heat storage technologies. In the case of multiple energy storage systems, energy management system is used. Another unit in a microgrid is the point of common coupling. It is the point in the electric circuit where a microgrid is connected to a main grid [10]. Figure 13.1 describes various microgrid components.

13.2.3 Challenges

Considering the reliability and full utilization of distributed generation units, the introduction of microgrids leads to a lot of operational challenges which have to be considered while designing the operation, control and protection systems. Some of the challenges are bidirectional power flows, like microgrids with distributed

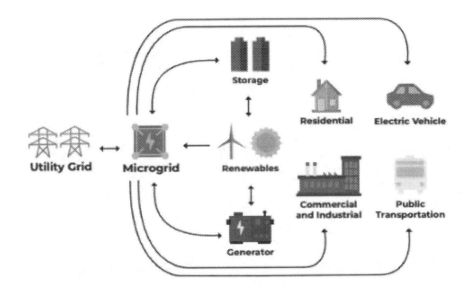

FIGURE 13.1 Interconnection of microgrid system.

generation sets causing reverse power flow, which results in disturbed power flow patterns, voltage problems and uneven distribution of fault current [11]. Issues related to stability, low inertia, modeling and uncertainty are some other challenges.

13.3 RENEWABLE ENERGY SOURCES: OVERVIEW

With the rise of energy consumption and growing demand for electric power every day, there is a continuous need for adequate generation and transmission facilities [12]. Environmental concerns about fossil fuels and their constraints have spurred great interest in generating electricity from non-conventional sources of energy [13]. Considering energy efficiency and power quality issues, the generation of electrical energy from renewable sources is an important factor for meeting world energy requirements using sources like solar, wind, geothermal, biomass and ocean energy [14]. Solar energy is predominant due to its free availability worldwide. Solar energy can be utilized in two ways: concentrated solar power and solar PV [15]. Solar PV system has various advantages in comparison to solar thermal. PV plants are mostly installed in the United States and Spain. Even India is also dominant among users of various renewable energy sources [16]. It has been estimated that electricity from PV generation will contribute 7% of world electricity needs by 2030, which will rise to 25% by the year 2050. PV is one of the fastest-growing technologies, with an annual growth rate of 35–40% [17].

The present installed power generation capacity of India is 334.40 GW as of January 31, 2018. Of the total installed capacity, 67.8% is conventional power plants, and 32.2% is non-conventional power plants. Of the total installed power capacity, renewable energy accounted for 18.37%, excluding large hydropower. The Solar

Energy Corporation of India is responsible for the development of solar energy industries in India. As of January 2018, installed solar power has reached over 20 GW, including both solar parks and rooftop solar panels. India is ranked number one taking into consideration solar energy produced per watt with an irradiance of 1700 to 1900 kW/hours per kilowatt peak (kWh/kWp).

13.3.1 Photovoltaic Standalone Systems

PV systems are either used as a standalone system or connected to the utility grid. In a standalone system, power from the PV array is directly fed to the load instead of connecting to the utility system [28]. In rural areas, standalone systems are an economic way of utilizing PV energy due to long-term availability of solar radiation and rare access to the utility grid [29]. The various applications of PV standalone systems are lighthouses, military applications, water pumping systems and communication systems [30]. Other than providing auxiliary power, the various drawbacks include cost and limited storage capacity (batteries) and as a result dissipation and waste of surplus energy generated. As the standalone system is not connected to the grid, there is always a requirement for a storage element during the off-peak demand period [31]. Another important feature of the standalone system is that there must be a balance between operational capacity and maximum load demand [32].

Nonconventional energy is one of the prime areas where more focus can be given to decrease the use of fossil fuel. Solar energy finds wide application, as it is clean, pollution free and available free of cost [18]. Studies show that the short-circuit current of the module decreases as the band gap increases. The optimum band gap for the ideal solar cell is 1.45 eV [19]. The photon-generated current is directly proportional to the intensity of solar irradiance. The short-circuit current has a logarithmic relation to irradiance. Fill factor is an indicator of the quality of the cell. Most commercially available silicon solar cells have a fill factor in the range 0 to 1 [20]. Some researcher designed two diode models and found better results. But the circuit becomes complex with the design. Solar irradiance is a positive factor, but studies also show that with an increase in irradiance, the total amount of energy is not converted to electricity; it gets converted to heat energy, thereby increasing the temperature of the module [21]. As a result, the module efficiency decreases at high temperatures as the band gap drops [22]. The output efficiency of PV modules depends on a variety of factors like solar irradiance, temperature, the presence of dust, shading, panel orientation, tracking, cell structure and so on [23].

13.3.2 Design of Simple DC Microgrid for AC Load

The system is designed on the MATLAB/Simulink platform. The system designed is shown in Figure 13.2 and consists of a PV array, which acts like a source converting solar energy into electrical energy, a DC-DC converter for carrying out load matching, a voltage source inverter to convert DC to AC voltage and an AC motor with centrifugal pump load. The AC motor used here is a permanent magnet synchronous motor. The system is tested with the proposed MPPT technique for better performance.

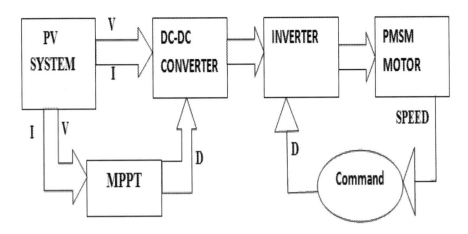

FIGURE 13.2 Block diagram of PV system connected for AC load.

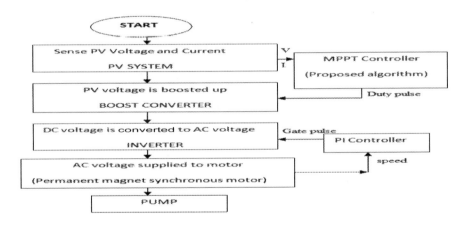

FIGURE 13.3 Flowchart of MPPT algorithm for PV water pumping system.

13.3.3 MATLAB/Simulink

The simulation is done in MATLAB/Simulink. A PV array is designed using six of a 180-Watt SANYO HIT solar module. Solar insolation and temperature are used as input to the system. In this work, standard conditions are taken into account: irradiance of 1,000 W/m² and temperature of 25°C. The PV system is connected to the boost converter. The MPPT controller provides the pulse to the converter. The simulation in Figure 13.4 consists of a PV array connected to the boost converter and voltage source PWM inverter, and the output of the inverter is connected to the permanent magnet synchronous motor. The switch to the power metal oxide semiconductor field effect transistor of the converter is given from the MPPT controller. Different MPPT techniques are used in the design under the same irradiance conditions.

Case Studies on Microgrid Design Using Renewable Energy 285

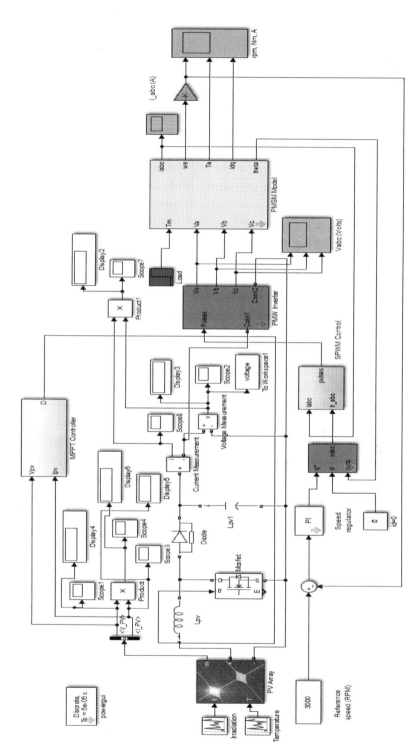

FIGURE 13.4 Simulink model of the designed system.

The simulation in Figure 13.4 consists of a PV array connected to the boost converter and resistive load. Input power, voltage and current and output power and voltage and current of the boost converter are measured.

13.3.4 Results and Discussion

The output of the MATLAB simulation carried out in this work is represented as follows.

Figures 13.5 and 13.6 show the I-V and P-V curves of the PV array at various irradiance levels: 1000, 800, 600, 400 and 200 W/m². From both figures it is observed that the maximum power point changes with the change in the irradiance. Figure 13.7

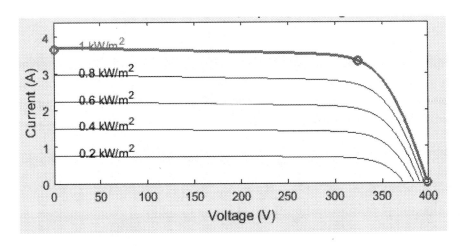

FIGURE 13.5 I-V curve of the PV array at various irradiances.

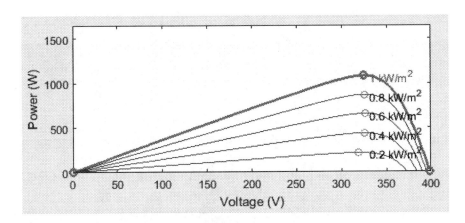

FIGURE 13.6 P-V curve of the PV array at various irradiances.

Case Studies on Microgrid Design Using Renewable Energy 287

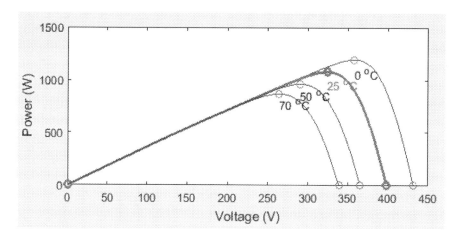

FIGURE 13.7 P-V curve of the PV array at various temperatures.

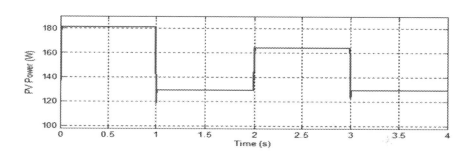

FIGURE 13.8 PV array output power.

shows the P-V curve of the PV array at various temperatures, that is, 75°C, 50°C, 25°C, 0°C. Even with the change in temperature level, the maximum power point changes. With the increase in voltage, power decreases, whereas with the increase in voltage, a marginal change in current is noticed.

Figure 13.8 shows the tracking of the maximum power point at various irradiance levels. The results show good performance under the steady-state condition. Around the peak point, the oscillations are mostly eliminated. The duty ratio is fixed once the maximum power point is reached.

The output voltage of the PV array that is input to the converter is 324 volts. The input current to the boost converter is 3.3 amperes. Thus, Figure 13.9 shows the power input to the converter, 1,080 kW.

Figure 13.10 shows the output power of the boost converter using different MPPT methods. From the figure, the proposed method gives the better result in comparison to other methods.

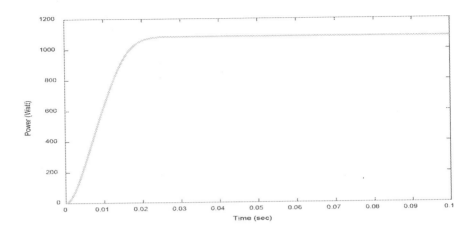

FIGURE 13.9 Input power of the converter.

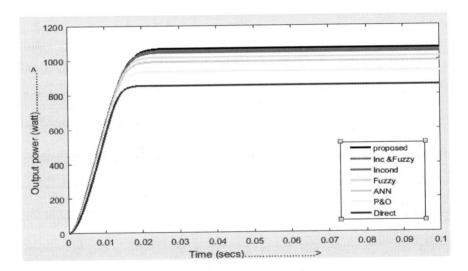

FIGURE 13.10 Output power of the converter using different MPPT methods.

Figure 13.11 shows the output line voltage of the inverter. The current level in the simulated line current waveform for PV VSI is 1.43 A.

From Figure 13.13, it is observed that the speed of the PMSM motor is around 3000 rpm. As it is a high-speed motor, the simulation result shows a value close to the set speed.

13.4 HYBRID SYSTEM USING HOMER SOFTWARE

HOMER is a optimization tool mainly meant for modeling the physical behavior of power systems and calculating lifecycle costs, including installation and operation

Case Studies on Microgrid Design Using Renewable Energy 289

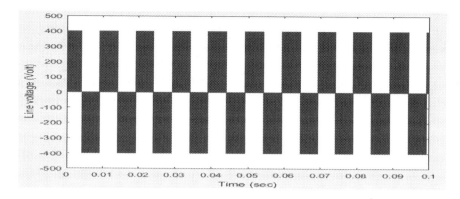

FIGURE 13.11 Output line voltage of the inverter.

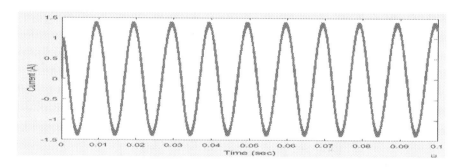

FIGURE 13.12 Output current of the inverter.

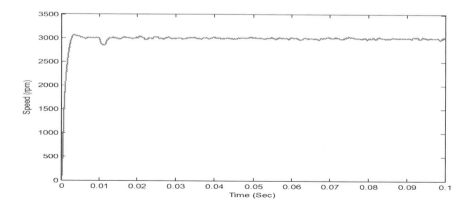

FIGURE 13.13 Speed of PMSM motor.

cost of the system. The modeler provides data on renewable energy, building load profile and cost components, which is fed to the HOMER. The software performs an hourly calculation of the energy balance for the proposed configuration for a fixed period of time. Sensitivity analysis can also be performed to check the effect of uncertainty. The hybrid system takes into account solar, wind, the diesel generator and the utility grid.

13.4.1 MICROGRID SYSTEM MODELING: CASE STUDY

Let us consider a standalone building at the Kalinga Institute of Technology, Bhubaneshwar. The building is connected to the utility grid. During grid outages, the diesel generator installed on campus can be used. Thus, we design here a microgrid architecture where the electric energy produced from solar and wind augments the grid supply to meet the demand. A converter is connected for conversion of DC power produced by the solar PV system to AC. A 394-V battery bank is connected to a DC bus for energy storage and used as backup energy source. The modeling of the proposed system is done by HOMER.

The microgrid system is a combination of different units like the PV, wind turbine, utility grid, diesel generator, converter and battery. Both technical and economic details of each unit are required for the simulation process. Along with these economic inputs, constraints and control parameters are to be considered when designing a microgrid in HOMER.

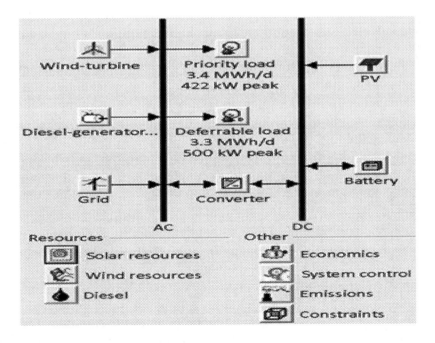

FIGURE 13.14 System model of microgrid.

13.4.1.1 PV Modeling

The PV array is considered in the 120 to 300 kW range. The cost of the PV panel is taken as $3,650/kW. The operation cost of the PV array is neglected, and the maintenance cost is also not considered. A lifetime of 20 years, is selected, with derating factor 82%. A dual-axis tracking system is considered for the PV panel.

13.4.1.2 Wind Turbine Modeling

The kinetic energy from the wind is converted to AC energy. A four-step process is followed for the calculation of power output from the wind turbine. Per wind resource data, the average wind speed for an hour is determined at anemometer height. Next, by using power laws, the wind speed at the turbine's hub height is calculated. Then power output is calculated at a particular wind speed by referring to the turbine's power curve and assuming air density to be at standard conditions. The turbine rating is considered 45 kW AC. Three turbines are considered. Each turbine's cost is around $181,042. The operation and maintenance cost is $4,256/year/turbine. This reduces as the number of turbines increases. The lifetime is considered 20 years with a hub height of 45 m.

13.4.1.3 Battery Modeling

Batteries are used for storage purposes and operate when there is a power deficit. Thus, a battery bank is designed for a 394-V DC bus. The rating of each battery is 12 V, 200 Ah and 2.4 kWh. A string of 32 series-connected batteries are used for power delivery at the desired voltage level. The battery cost is $490, and the replacement cost is $420. The lifetime is considered 12 years.

13.4.1.4 Diesel Generator Modeling

The diesel generator produces electricity and gives out heat as a byproduct by consuming fuel. These generators are used in microgrids to cover peak load. When renewable energy sources produce less power, not enough to meet the requirements, the diesel generator meets the electric load demand.

Three diesel generators are considered with different capacities: the first is 375 Kw and the second and third are 1060 Kw. The operation and maintenance cost is considered $0.003/hour, and the cost of diesel is around $0.9/ltr.

13.4.1.5 Converter Modeling

The converter works both as a rectifier and inverter, and the rating is considered per the total PV panel output. The converter converts PV output to AC. The lifetime of the converter is 25 years, with 90% efficiency both ways. The size is considered 130 to 300 kW at a rate of $120/kW and replacement cost $90/kW.

13.4.1.6 Utility Grid

The building considered produces a maximum demand of 650 kW on the utility grid. The electricity tariff is $0.065/kWh of energy consumption and the demand rate $6/Kw/month.

13.4.1.7 Load Profile

Various types of loads are connected to the building, like the uninterrupted power supply (UPS), heating, lighting, ventilation, air conditioning and other utility loads. The system categorizes the load smartly into two types, priority load and deferred load. Priority loads are the electrical demand which has to be served at a specific time, like the UPS and elevators. These types of load demands are not fixed and change from day to day depending on the user requirement. A deferrable load is a calculated load. One can easily track the nature and amount of demand for a certain period, like battery charging stations and water pump loads.

13.4.2 Energy Management

In this microgrid system, energy management is considered for three conditions: normal operating conditions, grid outage and total blackout. In design, it is assumed that the renewable energy source supplies the nominal power while the grid supplies the rated power [24]. In India, load shedding is a frequent problem and is also unpredictable; hence, it becomes a real challenge for the microgrid to both manage the system and make it reliable [25]. The normal operating condition is a state where energy from the renewable source is produced at a nominal rate and the grid supplies

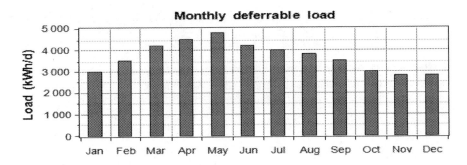

FIGURE 13.15 Monthly deferrable load profile for a year.

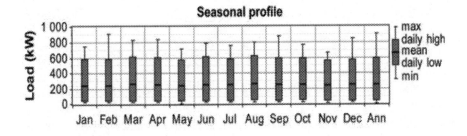

FIGURE 13.16 Monthly priority load profile for a year.

Case Studies on Microgrid Design Using Renewable Energy 293

the required energy demand; that is, the grid serves the priority loads, and renewable energy sources serve the deferrable loads [26]. Battery charging is done by the amount of excess energy from the RES source. Next, whenever there is a fault in the distribution and transmission line or during load shedding, that is, when the utility is not in a condition to serve the load, a grid outage condition occurs [27]. Thus, the renewable source supplies energy to priority loads like UPS, and the excess is fed to other priority loads. The third condition mentioned is total blackout, which is a state where an outage of both the renewable energy source and grid takes place simultaneously. This situation rarely occurs and is the condition of the worst-case scenario. Even during such a situation, the microgrid is designed to be capable of meeting the priority load. The diesel generator installed has to be synchronized whenever required. Both the battery and distributed generation set serve the priority load and can also serve the deferrable load if needed.

13.4.3 SIMULATION

The microgrid designed in this chapter is simulated considering various factors like load profile, component cost and size, control strategy and system economics. By using HOMER software various possible combinations and their feasibility are checked to form the Microgrid.

Figures 13.17, 13.18 and 13.19 show load variations of the building on an hourly basis and a comparison with PV and total renewable energy source output. Figure 13.20 describes the power generated from renewable energy sources on a monthly basis per year. Figures 13.21 and 13.22 show power purchased from and sold to the grid on yearly basis. Figure 13.23 shows the total electricity production by month from different renewable sources.

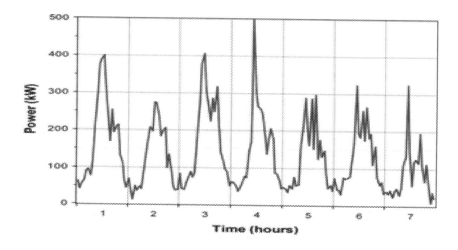

FIGURE 13.17 Load variation of building on hourly basis.

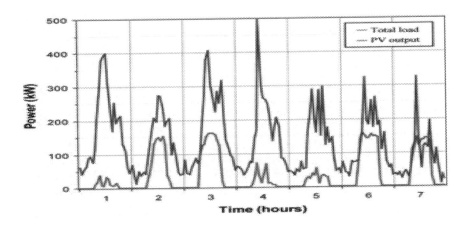

FIGURE 13.18 Variations of load on hourly basis versus PV output.

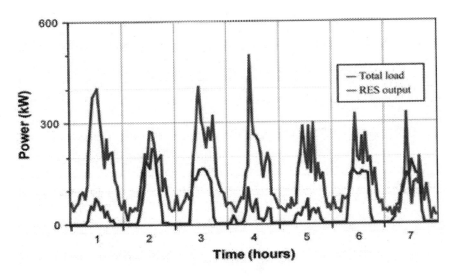

FIGURE 13.19 Variations of load on hourly basis versus total RES output.

Table 13.2 shows the optimization result for the designed microgrid system. The optimization result shows that the most economic system configuration is achieved for a combination of a 650-Kw utility grid, 200-kW PV and three units of wind turbine; 33 batteries; and a 180-kW converter. The total net present cost is $2878218, with a capital cost of $968531, replacement cost $27145, operating cost $149372, $0 fuel cost and salvage value $3,245. The best configuration does not consider a diesel generator for its operation.

Case Studies on Microgrid Design Using Renewable Energy 295

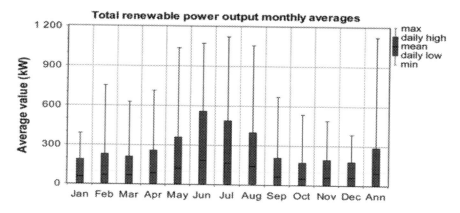

FIGURE 13.20 Month-wise power generation from renewable sources per year.

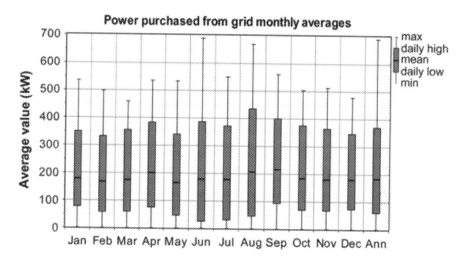

FIGURE 13.21 Monthly average of power purchased from grid.

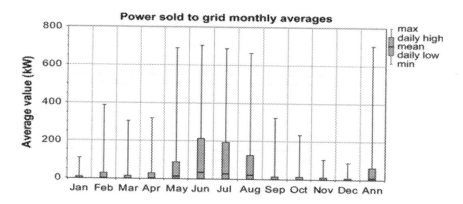

FIGURE 13.22 Monthly averages of power sold to grid in a year.

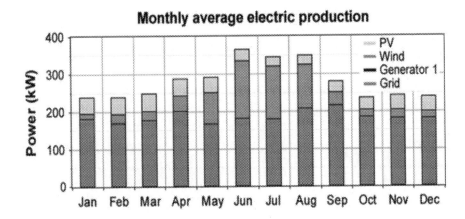

FIGURE 13.23 Electricity production month-wise from different sources.

TABLE 13.2
Optimization Results for the Microgrid System

PV (kw)	WT	Distributed generation (kW)	Battery	Converter (kW)	Grid (kW)	Initial capital	Operating cost ($/year)	Total NPC	COE ($/kWh)	Ren. fraction
200	3		33	180	650	968531	149372	2878218	0.094	0.36
180	3	375	33	170	650	947943	158834	2978484	0.098	0.34
200	3			180	650	956732	159504	2995782	0.098	0.37

Note: Indian numerical value points used for fractions. Necessary values are given in the table.

13.5 CONCLUSION

Microgrids are built of various hybrid interconnections, such as both alternate and direct energy sources, different storage systems, varieties of load and different types of converters. The architectural selection depends on geographical, economical and technical factors.

In this work, a PV array is simulated along with a boost converter and inverter, with a permanent synchronous motor used for water pumping application. A single-phase PWM inverter is used for the conversion of DC output from the PV to AC input to the motor. Several MPPT techniques are implemented to get the best duty cycle so that the output power obtained from the converter can be maximized. Using DC motors, inherent losses occurs at brushes, and the requirements of maintenance are also greater. Hence, the use of AC motors is simpler and more reliable. The speed of the motor is around 3,000 rpm, and it can be used for high-speed operation. The design of the water pumping system helps rural people store water in storage tanks which can be used both for irrigation and other domestic purposes.

In the second case study for a hybrid microgrid system, HOMER provides us with the microgrid design and best optimization result. The building considered for the design had a priority load of 3.2 MWh/day and deferrable load of 3.1 MWh/day. The optimization result shows that the most economic system configuration is achieved for combination of 650-Kw utility grid, 200-kW PV and three units of wind turbine; 33 batteries; and a 180-kW converter. The system designed is environmentally friendly, which leads to 6.23% annual cost savings and a 36.4% reduction in carbon dioxide emissions. Even under various environmental conditions, like solar radiation, wind speed and changes in energy consumption patterns, the system runs well and provides a reliable supply.

REFERENCES

1. National Action Plan on Climate Change, *Ministry of Environment, Forest, and Climate Change*. Government of India. http://envfor.nic.in/ccd-napcc.
2. K. T. Akindeji, I. E. Davidson, "Microgrid and Active Management of Distribution Networks with Renewable Energy Sources," Proceedings of the 24th South African Universities Power Engineering Conference (SAUPEC), Vanderbijlpark, South Africa, 26–28 January 2016, pp. 171–175.
3. K. T. Akindeji, "A Review on Protection and Control of Microgrid with Distributed Energy Resources Integration," Presented at Protection, Automation and Control (PAC) World Conference, University of Strathclyde, Glasgow, Scotland, 29 June–2 July 2015.
4. H. Zhou, T. Bhattacharya, D. Tran, T. S. T. Siew, A. M. Khambadkone, "Composite Energy Storage System Involving Battery and Ultracapacitor with Dynamic Energy Management in Microgrid Applications," *IEEE Transactions on Power Electronics*, Vol 26 No 3, pp. 923–930, 2011.
5. A. O. Mafimidiwo, "Impact of Three-Dimensional Photovoltaic Structure on Solar Power Generation," PhD thesis, School of Engineering, University of KwaZulu-Natal., Durban, South Africa, 2016.
6. www.solargis.com/maps-and-gis-data/download/south-africa.
7. L. Hadjidemetriou, et al., "Design Factors for Developing a University Campus Microgrid," 2018 IEEE International Energy Conference (ENERGYCON), Limassol, pp. 1–6, 2018.
8. B. K. Bose, "Neural Network Applications in Power Electronics and Motor Drives—An Introduction and Perspective," *IEEE Transactions on Industrial Electronics*, Vol 54, pp. 14–33, 2007.
9. M. Veerachary, et al., "Neural-Network-Based Maximum-Power-Point Tracking of Coupled-Inductor Interleaved-Boost-Converter-Supplied PV System Using a Fuzzy Controller," *IEEE Transactions on Industrial Electronics*, Vol 50, pp. 749–758, 2003.
10. T. Noergaard, *Embedded Technology: Embedded Systems Architecture: A Comprehensive Guide for Engineers and Programmers*. Burlington, MA: Newnes. ProQuest Library, 2005.
11. S. S. Chandel, et al., "Review of Solar Photovoltaic Water Pumping System Technology for Irrigation and Community Drinking Water Supplies," *Renewable and Sustainable Energy Reviews*, Vol 49, pp. 1084–1099, 2015.
12. Joao Victor Mapurunga Caracas, Guilherme de Carvalho Farias, Luis Felipe Moreira Teixeira, Luiz Antonio de Souza Ribeiro, "Implementation of a High Efficiency, High Lifetime, and Low-Cost Converter for an Autonomous Photovoltaic Water Pumping System," *IEEE Transactions on Industrial Applications*, Vol 50 No 1, January–February 2014.

13. A. B. C. S. B. Slama, A. Chrif, "Efficient Design of a Hybrid (PV-FC) Water Pumping System with Separate Mppt Control Algorithm," *IJCSNS International Journal of Computer Science and Network Security*, Vol 12, pp. 53–60, January 2012.
14. M. A. Vitorino, M. B. R. Correa, C. B. Jacobina, A. M. N. Lima, "An Effective Induction Motor Control for Photovoltaic Pumping," *IEEE Transactions on Industrial Electronics*, Vol 58 No 4, pp. 1162–1170, April 2011.
15. H. Harsono, "Photovoltaic Water Pump System," Ph.D. dissertation, Dept. Intell. Mech. Syst. Eng., Faculty Kochi Univ. Technol., Kochi, Japan, August 2003.
16. M. Chunting, M.B.R. Correa, J.O.P. Pinto, "The IEEE 2011 International Future Energy Challenge-Request for Proposals," *Proc. IFEC*, pp. 1–24, 2010.
17. J. S. Ramos, H. M. Ramos, "Solar Powered Pumps to Supply Water for Rural or Isolated Zones: A Case Study," *Energy for Sustainable Development*, Vol 13, pp. 151–158, 2009.
18. T. P. Correa, S. I. Seleme Jr., S. R. Silva, "Efficiency Optimization in the Stand-Alone Photovoltaic Pumping System," *Renewable Energy*, Vol 41, pp. 220–226, May 2012.
19. V. Badescu, "Dynamic Model of a Complex System Including PV Cells, Electric Battery, Electrical Motor, and Water Pump," *Energy*, Vol 28, pp. 1165–1181, 2003.
20. Bo Yuan, Xu Yang, Xiangjun Zeng, Jason Duan, Jerry Zhai, Donghao Li, "Analysis and Design of a High Step-Up Current-Fed Multi Resonant DC-DC Converter with Low Circulating Energy and Zero-Current Switching for All Active Switches," *IEEE Transactions on Industrial Electronics*, Vol 59 No 2, February 2012.
21. R. Chenni, L. Zarour, A. Bouzid, T. Kerbach, "Comparative Study of Photovoltaic Pumping Systems Using a Permanent Magnet Synchronous Motor (PMSM) and an Asynchronous Motor (ASM)," *Rev. Energy. Ren*, Vol 9, pp. 17–28, 2006.
22. L. Zarour, R. Chenni, A. Borni, A. Bouzid, "Improvement of Synchronous and Asynchronous Motor Drive Systems Supplied by Photovoltaic Arrays with Frequency Control," *Journal of Electrical Engineering*, Vol 59 No 4, pp. 169–177, 2008.
23. R. Chenni, L. Zarour, A. Bouzid, T. Kerbach, "Comparative Study of Photovoltaic Pumping Systems Using a Permanent Magnet Synchronous Motor (PMSM) and an Asynchronous Motor (ASM)," *Rev. Energy. Ren*, Vol 9, pp. 17–28, 2006.
24. M. Dubey, S. Sharma, R. Saxena, "Solar PV Standalone Water Pumping System Employing Pmsm Drive," Electrical Electronics and Computer Science (SCEECS), 2014 IEEE Students Conference on pp. 1–6, March 2014.
25. C. Jin, P. Wang, J. F. Xiao, et al., "Implementation of Hierarchical Control in DC Microgrids," *IEEE Trans Ind Electron*, Vol 61 No 8, pp. 4032–4042, 2014.
26. C. Cristea, J. P. Lopes, M. Eremia, et al., "The Control of Isolated Power Systems with Wind Generation," Proceedings of the 2007 IEEE Lausanne Power Technology Conference, Lausanne, pp. 567–572, 1–5 July 2007.
27. A. Mohd, E. Ortjohann, W. Sinsukthavorn, et al., "Supervisory Control and Energy Management of an Inverter-Based Modular Smart Grid," Proceedings of the IEEE PES Power Systems Conference and Exposition (PSCE'09), Seattle, p. 6, 15–18 March 2009.
28. Arjyadhara Pradhan, Bhagabat Panda, "Performance Analysis of Photovoltaic Module at Changing Environmental Condition Using MATLAB/Simulink," *International Journal of Applied Engineering Research*, Vol 12 No 13, pp. 3677–3683, 2017.
29. Arjyadhara Pradhan, Bhagabat Panda, "Experimental Analysis of Factors Affecting the Power Output of the PV Module," *International Journal of Electrical and Computer Engineering*, Vol 7, pp. 141–149, 2017.
30. Arjyadhara Pradhan, Bhagbat Panda, "Design and Modelling of Simplified Boost Converter for Photovoltaic System," *International Journal of Electrical and Computer Engineering*, Vol 8 No 1, pp. 132–139, 2018.

31. Arjyadhara Pradhan, Bhagabat Panda, "Analysis of Ten External Factors Affecting the Performance of PV System," International Conference on Energy, Communication, Data Analytics and Soft Computing (ICECDS 2017), IEEE Madras Section 1st–2nd August, SKR Engineering College, Tamil Nadu.
32. Arjyadhara Pradhan, Bhagabat Panda, "Design of DC-DC Converter for Load Matching in Case of PV System," International Conference on Energy, Communication, Data Analytics and Soft Computing (ICECDS 2017), IEEE Madras Section 1st–2nd August, SKR Engineering College, Tamil Nadu.

Index

A

AC microgrid, 36, 45–46, 122, *123*, 238
adaptive protection, 248–249

B

benefit cost ratio (BCR), 107, 113

C

central generation, 25
centralized control system, 38–39, *39*
centralized grid systems, 23
chemical technologies, 33
communication delay, 84, 85, 88
compromised solution, 110
conclusion, 48, 78, 98, 114, 132, 177, 197, 212, 229, 250, 275, 296
controlled unit, 232
conventional grid, 21, 281
cost optimization, 24, 281
cumulative current price (CCP), 106
current, 2, 4, 9, 14, 22, 41, 56, 57, 61–66, 70–71, 77–78, 103–104, 123, 202–204, 208–209, 212, 218, 221–225, 234, 238–250, 276, 283, 286–288
customer microgrid, 43

D

data clustering, 106, 112, 116
DC microgrid, 36, 45–47, *46–47*, 122, *123*, 234, 238, *239*, 283
demand response (DR), 138–139, 147–148, 150
DER elements, 235–238
description of problem and methodology, 144–149
DIRECT, 15
distributed control system, 37, 39–40
distributed generation (DG), 21, 26–28, 32–39, 41–42, 44–46, 54, 102, 123, 125–127, 139, 184, 202, 232, 238–243, 245–246, 249–251, 280–281, 293
distribution network, 16, 22, 25, 27, 128, 220
dynamic crowding distance, 109–110, 113

E

electrical technologies, 32
electric grid, 20, 24, 28

energy management, 292–293
energy storage, 121–122
energy storage technology, 31–33

F

fault detection, 5, 46, 122
filter delay, 88–89
frequency regulation, 86
fuel cells, 27, 30–31, 102, 132, 182, 185
fuzzy logic, **12**, 110–111, 113, 182, 218
fuzzy-neuro hybrid forecasting, **12**

G

grid connected mode, 32–34, 42–43, 85–86, 93, 102, 121, 139

H

harmonics, 7, 21, 48, 125, 131, 202–203, 208, 212, 215, 228
hierarchical and distributed control, 85, 89, *91*
HOMER, 14, 138, 139, 290, 297
HOMER-grid software, 140, 144–147, 150, **159**, 177, 178, 293
hybrid microgrid, 45, 106, 138–139, 184, 297
hybrid system, 76, 102, 140, 147, 288–290

I

IHOGA, 14
inertia, 78, 84, 86, 282
interface module, 36, 41–43, 46
introduction, 2, 19, 54, 83, 102, 119, 138, 181, 201, 217, 232, 255, 280
inverter delay, 88–89
irradiation, 85, 88, 93–94, 141, *146*
islanded microgrids, 2–3, **5**, 6, 10, 14, 132
islanded mode, 6, 14, 25, 32–35, 37–38, 42–43, 85–86, 93–95, 98, 102, 121–122, 124–126, 131–132, 202, 232, 236, 238, 240, 242, 246
islanding mode, 85, 251

L

load, 4, 6, 25, 33, 42–46, 54, 86, 103, 111–112, 114, 143–144, 235, 250, 256–257, 276, 292
load disturbances, 93–94
localized control system, 37–38, 40

301

Index

M

macro grid, 233
MATLAB, 62, 64, 67, *68*, 78, 93, 140, 144–145, 189, 203–204, 206, 220, 226, 262, 283–284, 286
maximum power point tracking (MPPT), 64, 66–70, 78–79, 204, 237, 283–284, *284*, 287
mechanical technologies, 32
microgrid (MG), 13–14, 83–89, 93–95, 98, 201–203, 205, 211–212
microgrid central controllers (MGCC), 38–39, 127
microgrid design, 10, *10*, 146, 150, 297
microgrid loads, 33
minimum detail length, 248
modeling of microgrid system, 140–143
multi agent, 246–247, *247*
multi-objective optimization, 14, 103, 110, 114
multipurpose, 246, 252

N

non-dominant sort, 109

O

one-way traffic, **232**

P

parental inheritance, 109
Pareto, 109–111, 113–114, **116**, *116*, 139
point of common coupling (PCC), 2, 26–27, 34, *34*, 42, 208, 234, 281
pollutant emissions, 149, 157, **157**, 169, **170**, 177–178
power grid, 20, 27, 56–57, 59, 74, 121, 280
power quality, 202
primary control, 84, 89
probabilistic model, 103
programmable logic controller (PLC), 13
protection technique, 231–252
PWM, 71, 208, 284, 296

R

references, 17, 48, 79, 99, 117, 132, 178, 198, 215, 229, 252, 276, 297
reliability, 24, 32, 35, 44, 60, 102–103, 108, 138–140, 147–148, 167, 169, 178, 220, 229, 237
reliability of system, 147

renewable sources, 24, 31–32, 54, 102–103, 113, 122, 138–139, 149, 182, 280–282, 293
RES, 124–125, 256–257, 293, *294*
resilience of system, 148
results and discussion, 150, 210, 262, 286

S

salvage, 106, 162, 294
secondary control, 83–84, 130
security constraints, 103–108, 114
series transformer and the shunt APF (SAPF), 202, 208–209
short-circuit, 235, 242, 283
simulation studies, 93–98
solar power resource assessment for case study, 140
solar PV cells, 29
standalone reciprocating engine generators, 36
supervisory remote control unit (SRCU), 249
synchronous reference frame (SRF), 202

T

tertiary control, 84, 98, 130
Thevenin's law, 241
topology, 234–235
total net present cost and levelized cost of energy, 148–149
transfer function, 185–186
transient stability, 37, 183
transport delay, 89, 92, 94–95, *96*, 98
turbine generators, 3, 28–29

U

uncertainty, 102–106, 108, 114, 282, 290
unified power quality conditioner (UPQC), 202
utility microgrid, 33–34, 36–37, 42–45, 54, 76, 150, 201–202, 240, 291
utility tariff schemes, 150, *151*, 177

V

virtual microgrid, 43

W

wavelet packet transformation (WPT), 248
wind speed disturbance, 93
wind turbine(s), 8, 21, 30, 54–63, 76–78, 290
wind velocity, 87

Printed in the United States
by Baker & Taylor Publisher Services